Andreas Wingen

Stable and unstable manifolds in open chaotic systems

Andreas Wingen

Stable and unstable manifolds in open chaotic systems

With application to the tokamak TEXTOR-DED

Südwestdeutscher Verlag für Hochschulschriften

Impressum/Imprint (nur für Deutschland/ only for Germany)
Bibliografische Information der Deutschen Nationalbibliothek: Die Deutsche Nationalbibliothek
verzeichnet diese Publikation in der Deutschen Nationalbibliografie; detaillierte bibliografische
Daten sind im Internet über http://dnb.d-nb.de abrufbar.
Alle in diesem Buch genannten Marken und Produktnamen unterliegen warenzeichen-, marken-
oder patentrechtlichem Schutz bzw. sind Warenzeichen oder eingetragene Warenzeichen der
jeweiligen Inhaber. Die Wiedergabe von Marken, Produktnamen, Gebrauchsnamen,
Handelsnamen, Warenbezeichnungen u.s.w. in diesem Werk berechtigt auch ohne besondere
Kennzeichnung nicht zu der Annahme, dass solche Namen im Sinne der Warenzeichen- und
Markenschutzgesetzgebung als frei zu betrachten wären und daher von jedermann benutzt
werden dürften.

Verlag: Südwestdeutscher Verlag für Hochschulschriften Aktiengesellschaft & Co. KG
Dudweiler Landstr. 99, 66123 Saarbrücken, Deutschland
Telefon +49 681 37 20 271-1, Telefax +49 681 37 20 271-0, Email: info@svh-verlag.de
Zugl.: Düsseldorf, Heinrich-Heine Universität, Dissertation, 2006

Herstellung in Deutschland:
Schaltungsdienst Lange o.H.G., Berlin
Books on Demand GmbH, Norderstedt
Reha GmbH, Saarbrücken
Amazon Distribution GmbH, Leipzig
ISBN: 978-3-8381-0620-5

Imprint (only for USA, GB)
Bibliographic information published by the Deutsche Nationalbibliothek: The Deutsche
Nationalbibliothek lists this publication in the Deutsche Nationalbibliografie; detailed
bibliographic data are available in the Internet at http://dnb.d-nb.de.
Any brand names and product names mentioned in this book are subject to trademark, brand or
patent protection and are trademarks or registered trademarks of their respective holders. The
use of brand names, product names, common names, trade names, product descriptions etc.
even without a particular marking in this works is in no way to be construed to mean that such
names may be regarded as unrestricted in respect of trademark and brand protection legislation
and could thus be used by anyone.

Publisher:
Südwestdeutscher Verlag für Hochschulschriften Aktiengesellschaft & Co. KG
Dudweiler Landstr. 99, 66123 Saarbrücken, Germany
Phone +49 681 37 20 271-1, Fax +49 681 37 20 271-0, Email: info@svh-verlag.de

Copyright © 2009 by the author and Südwestdeutscher Verlag für Hochschulschriften
Aktiengesellschaft & Co. KG and licensors
All rights reserved. Saarbrücken 2009

Printed in the U.S.A.
Printed in the U.K. by (see last page)
ISBN: 978-3-8381-0620-5

Preface

At the beginning I would like to thank everyone who gave support for this dissertation.

I give my thanks to Prof. Dr. K.H. Spatschek for the selection of the topic and the detailed, intensive supervision during my thesis. Prof. Spatschek motivated me consistently by his lectures and many interesting discussions which kept my thesis going. It is due to him that I chose to become a theoretical physicist. Even today we are associated in an amicable and trusty employment relationship. I also give my thanks to Prof. Dr. A. Pukhov, who acted as the second reviewer for my diploma thesis as well, for being the second reviewer of my dissertation. Especially I want to thank Dr. S. Abdullaev from the Forschungszentrum Jülich, Germany, who always had plenty of time for discussions and came up with new suggestions.

I thank Dr. E.W. Laedke for providing his help with numerical problems and many helpful comments. Many thanks to Dr. M. Jakubowski from the Forschungszentrum Jülich, who helped relating the numerical results to measurements and experimental observations. I give my thanks to Dr. K.H. Finken, the former head of the TEXTOR experiment in Jülich, and Dr. T.E. Evans from General Atomics in San Diego, USA, who provided some interesting suggestions during several conferences. In the meantime Dr. Evans and I are also associated in a close employment relationship, especially since he offered me the opportunity to work with him in San Diego. For that and his continuing support he has my gratitude.

I also want to thank my friends and colleagues at the institute. In all the years of our common work, they always offered their help and support, especially Dr. G. Lehmann and Dr. Ch. Karle. This includes Dr. H. Wenk and E. Zügge as well, who took care (and E. Zügge still does) of the computer systems, and E. Gröters, who handles all organizational problems. All other members of the institute are included too, who create a friendly and supporting working atmosphere, which makes working at the institute for theoretical physics, Düsseldorf Germany, convenient.

Finally, I want to give my gratitude to my wife Andrea in particular. Her support made my academic studies and this thesis possible in the present form. Her love and understanding give me new strength every day. She invested countless hours checking this thesis' spelling. I dedicate this book to her!

Contents

1 Introduction   5

2 Formation of chaos in the symmetric tokamap regime   9
    2.1 The tokamap 10
    2.2 The mapping technique 12
    2.3 The tokamap in its symmetric form 15
    2.4 Statistical properties of the tokamap 17
    2.5 Construction of stable und unstable manifolds 21
    2.6 Periodic points 22
    2.7 Stable and unstable manifolds of the symmetric tokamap 25
    2.8 The question of spontaneously inverted q-profiles 27
    2.9 The symmetric revtokamap 28

3 Cylindrical model for magnetic field lines in TEXTOR-DED   33
    3.1 The current density 35
    3.2 The magnetic field of the DED coils 38
    3.3 The safety factor 43
    3.4 Hamiltonian of the DED field 46
    3.5 The DED map 48
    3.6 Characterization of the DED system by its statistical properties 52
    3.7 Topology of the stochastic edge region, analyzed by the stable and unstable manifolds 58

4 Toroidal DED model with relativistic particle drift effects   61
    4.1 The Hamiltonian for charged relativistic particles in an EM field 62
    4.2 Guiding-center approximation 64
    4.3 Simplification of the guiding-center equations 67
    4.4 The equilibrium field 69
    4.5 Explicit solution for ρ 71
    4.6 The perturbation field 72

4.7	The mapping procedure for the relativistic drift model	72
4.8	Unperturbed drift surfaces with varying kinetic energy	75
4.9	Drift effects with the DED perturbation field	80
4.10	Escape rates of particles and field lines	84
4.11	Non-relativistic Limit	87
4.12	Heat flux patterns in TEXTOR	89

5 Summary and conclusion   97

Appendix   103

   A   Discrete Schrödinger map 103
   B   Applying the Poisson summation rule 107
   C   Equivalent form of the DED map's generating function 109

Bibliography   111

Chapter 1

Introduction

The energy demand and consumption of our society has increased considerably in the past decades and it will increase even stronger in the next ones. A clean and efficient power source, which is independent of fossil fuels, is needed to secure the energy supply for the future. Fusion is a very promising candidate. One kilogram of Deuterium-Tritium fuel would release about 10^8 kWh of energy, which is sufficient for one day of operation of an 1 GW power plant. The fusion reaction produces Helium, which is the only waste product except of some radioactive isotopes in the containment structure, created by energetic neutrons. But compared to the waste products of nuclear fission, these isotopes have very small half-life periods.

The fusion reaction can only take place within high temperature plasmas at temperature levels of about $10-20$ keV. Further on, the Lawson criterion [1] has to be fulfilled. It demands for the energy confinement time τ_E: $nT\tau_E \geq 3 \times 10^{21}$ m^{-3} keV s, which is necessary to get stable fusion reactions inside the plasma at the given temperature and typical densities of about $n = 10^{14}$ cm^{-3}. Due to this demand, a good plasma confinement is needed. The tokamak, which uses magnetic plasma confinement, created by a strong helical field of about $2-4$ T, is one of the most advanced concepts. There are also other approaches like stellerators or laser induced fusion, but we will concentrate on the tokamak regime here.

On the other hand, the confinement of the plasma has to be controllable in such a way that the Helium can be removed from the plasma, while new Deuterium and Tritium is applied. Also the heat and particle deposition at the wall has to be regulatable to extract the excess energy from the plasma as well as to protect the wall from overheating and destruction. For this purpose the dynamic ergodic divertor, DED, was developed [2, 3] for the tokamak TEXTOR, Torus EXperiment for Technology Oriented Research, at the Forschungszentrum Jülich. The DED is a system of 16 helical coils, which creates a stochastic magnetic perturbation field at the edge of the tokamak [4, 5]. The DED field is a special external perturbation, but stochastic fields are also created by error fields, caused by misalignments within the geometry or the main coils. Therefore, stochastic magnetic fields are present in almost all fusion machines.

Chapter 1. Introduction

Additionally, special designed external coil systems are often used to directly influence particle motion to a wall of a fusion machine, to indirectly influence the heat and particle loads via suppression of ELMs, edge localized modes, or to change the plasma parameters in fusion plasmas like temperature profiles, toroidal rotation, etc [6, 7]. This makes the analysis of stochastic magnetic fields, the corresponding transport mechanisms (in principle externally produced) and the created wall patterns a highly relevant topic to nuclear fusion.

In this thesis, we discuss the statistics of stochastic magnetic fields by using discrete maps. As it is well known, magnetic field lines represent a $1\frac{1}{2}$-dimensional continuous Hamiltonian system. The main goal of mapping models is to replace the original continuous dynamical system, the magnetic field lines, by a discrete iterative map, which runs much faster then the small-step numerical integration [4, 8, 9, 10]. Mappings should be symplectic (or flux-preserving). They should have the same periodic points as the Poincaré map of the original system, and they should show the same regular and chaotic regions as the continuous magnetic field line evaluation. For global maps, a magnetic axis should be mapped to itself, and the magnetic flux should be always positive [11]. Thus, the transition to useful discrete maps is by no means trivial. We will use the mapping technique in its symmetrical form as derived by S. Abdullaev et al. [4, 5, 7, 8, 9, 10].

On the basis of the Hamiltonian mapping technique, presented in Sec. 2.2, we will study the transport mechanisms of heat and particles in stochastic fusion plasmas. We will analyze the chaotic motion inside the plasma as well as the wall patterns, created by open chaotic systems like the DED at TEXTOR. This leads to the following main questions, this thesis is dealing with:

- How are chaotic layers formed?

- What are the mechanisms of chaotic transport in stochastic magnetic fields?

- What are the transport mechanisms of heat and particles to the wall in open chaotic systems like the TEXTOR-DED?

- How are the structures of the wall patterns formed and what is the dominant influence?

- What are additional particle effects, especially for high energetic particles?

In order to get a better understanding of stochastic magnetic fields and their dynamics, regarding the questions above in particular, we will use the concept of the stable and unstable manifolds of unstable objects such as hyperbolic periodic points [12]. Many nonlinear phenomena can be explained by understanding the behavior of the unstable dynamical objects, which are present in the dynamics. The stable and unstable manifolds of hyperbolic points have recently attracted the attention of groups dealing with stochastic magnetic fields in tokamak fusion machines, especially the group of Todd Evans, who deals with the stable and unstable manifolds in the DIII-D [13, 14].

Due to the enormous relevance of these manifolds, we will study their structures and dynamics in several steps. First we will use a basic model to get a better understanding of their connections to chaos in principle. Then we will shift to a simplified realistic model, which can still be calculated analytically: the cylindrical DED model. At the end we will generalize to the real toroidal TEXTOR-DED model, including particle drift effects.

To investigate the topics above, following the outlined path, this thesis is organized as follows: In the first chapter the symmetric tokamap will be derived from a continuous Hamiltonian system, using the mapping technique, to principally determine the relation between the stable and unstable manifolds and the formation of chaotic layers and the transport within. We will compare the tokamap in its symmetric form with the non-symmetric one [11, 15, 16, 17], originally proposed by Balescu [11], by analyzing their statistical properties. We will show that, in addition to common properties, there are some fundamental differences. The structures and the effects of the stable and unstable manifolds will be studied for the symmetric tokamap model with monotonic and non-monotonic, reversed shear, q-profile [18]. The latter corresponds to the revtokamap [15]. In the next chapter we will shift to a more realistic model, a simplified cylindrical model for magnetic fiel lines in TEXTOR-DED, which can be calculated totally analytical. The magnetic field and the Hamiltonian for the DED system will be derived [5]. We will characterize the structures of the open chaotic DED system by its statistical properties, especially by the mean square displacement and the Lyapunov exponents. Considering the stable and unstable manifolds and the results from the tokamap regime, the transport, especially to the wall, will be analyzed for the magnetic field lines. We will show the typical structures of the DED system and their connection to the stable and unstable manifolds. In chapter 4 we will extend the DED model to the real toroidal geometry and include relativistic particle drift effects. Starting with the relativistic Hamiltonian for a particle in an electromagnetic field, the Hamiltonian for the guiding-center will be derived and a 4-dimensional mapping procedure for explicit time depending particle drifts will be constructed. The drift effects will be studied for the unperturbed and the perturbed case. The latter includes the stochastic DED field. We will show the non-relativistic limit, to compare with [19] and [20]. At the end we will calculate, analyze and explain real measurements of heat flux patterns at the divertor plates of the TEXTOR-DED theoretically, using the stable and unstable manifolds.

Chapter 1. Introduction

Chapter 2

Formation of chaos in the symmetric tokamap regime

In order to get a better understanding of the complex dynamics within chaotic layers, especially transport mechanisms, we first have to investigate how chaotic layers are formed, how the chaotic layers of different island chains act on each other and how the chaotic dynamics, especially the transport, change with changing perturbation strengths. The stable and unstable manifolds of hyperbolic periodic points are a new and very promising concept. So first we have to analyze the connection of the stable and unstable manifolds to the chaotic layers and their dynamics in principal. Many interesting mappings have been proposed [4, 11, 21, 22]. Discrete maps are often used to reduce the description of dynamical systems to algebraic iteration procedures. The standard map [23] is for example well known, which describes the dynamics of a kicked rotator. Another interesting map is the discrete Schrödinger map, see Appendix A, which reduces the discrete nonlinear Schrödinger equation [24] to a generalized form of the standard map. As shown in the appendix, this map can be used to construct solitary solutions of the discrete nonlinear Schrödinger equation.

One of the basic models for chaotic dynamics in tokamaks is the so called tokamap model [11, 15, 16, 17]. The tokamap was proposed by Balescu et al. [11] to describe the global behavior of magnetic field lines in tokamaks. It was constructed as an iterative discrete map, which is compatible with the toroidal geometry. Balescu et al. [11] presupposed that it represents the Poincaré map of a continuous magnetic field line system. For a Poincaré map, a toroidal cross section $\varphi_k = 2\pi k$ mod 2π of the tokamak is chosen. Then, at the intersection of the magnetic field line, the toroidal flux coordinate ψ (being proportional to the minor radius of the torus) is plotted against the poloidal angle θ mod 2π.

Chapter 2. Formation of chaos in the symmetric tokamap regime

2.1 The tokamap

Magnetic field lines are three-dimensional curves, $\vec{x}(s) = (x(s), y(s), z(s))$, tangent to a magnetic field $\vec{B}$, determined by a set of equations, $d\vec{x}/ds = \vec{B}$, where $ds = |\vec{B}|^{-1}dl$ is related to the element of length along a field line, i.e. $dl = (dx^2 + dy^2 + dz^2)^{1/2}$.

Ideally, in magnetically confined plasmas such as tokamaks and stellarators, magnetic field lines are lying on nested toroidal surfaces (magnetic surfaces) wound around a circular closed magnetic field line, the so called magnetic axis. The toroidal flux ψ is defined as the magnetic flux through the section of magnetic surface perpendicularly to the magnetic axis ($\psi = 0$). The field line coordinate on the magnetic surface is uniquely determined by a poloidal angle θ (the short way around the torus) and a toroidal angle φ (the long way around the torus). We use the magnetic flux ψ normalized to its value at the plasma boundary, i.e. $\psi = 1$ is at the plasma boundary. The divergence-free magnetic field $\vec{B}$ can be always presented in the Clebsch form [25] using these variables

$$\vec{B} = \nabla\psi \times \nabla\theta - \nabla H \times \nabla\varphi . \qquad (2.1.1)$$

Using the total differentials, which can be calculated by the directional derivatives

$$d\psi = \vec{B} \cdot \nabla\psi = -(\nabla H \times \nabla\varphi) \cdot \nabla\psi = -(\frac{\partial H}{\partial \theta}\nabla\theta \times \nabla\varphi) \cdot \nabla\psi = -\frac{\partial H}{\partial \theta}(\nabla\psi \times \nabla\theta) \cdot \nabla\varphi$$

$$d\theta = \vec{B} \cdot \nabla\theta = -(\nabla H \times \nabla\varphi) \cdot \nabla\theta = -(\frac{\partial H}{\partial \psi}\nabla\psi \times \nabla\varphi) \cdot \nabla\theta = \frac{\partial H}{\partial \psi}(\nabla\psi \times \nabla\theta) \cdot \nabla\varphi$$

$$d\varphi = \vec{B} \cdot \nabla\varphi = (\nabla\psi \times \nabla\theta) \cdot \nabla\varphi ,$$

one can derive equations for the magnetic field lines, which are of Hamiltonian form

$$\frac{d\psi}{d\varphi} = -\frac{\partial H}{\partial \theta}, \qquad \frac{d\theta}{d\varphi} = \frac{\partial H}{\partial \psi}, \qquad (2.1.2)$$

where a pair of variables (θ, ψ) represents canonical variables, and the toroidal angle φ plays the role of a time-like variable. The Hamiltonian $H = H(\psi, \theta, \varphi)$ is the poloidal flux. Using the Clebsch form, we can always derive a Hamiltonian description of magnetic fields. The equilibrium magnetic field configuration with nested magnetic surfaces $\psi(x, y, z) = $ constant follows from a "time"– independent Hamiltonian $H = H(\psi)$. Then, the field line equations (2.1.2) are integrable: $\psi = $ constant, $\theta = (\varphi - \varphi_0)/q(\psi)$, where $q(\psi) = 1/\frac{\partial H_0(\psi)}{\partial \psi}$ is known as the safety factor.

In the presence of non-axisymmetric magnetic perturbations, the poloidal flux H can be presented as a sum of the unperturbed flux $H_0(\psi)$ and the perturbed part $\varepsilon H_1 = \varepsilon H_1(\psi, \theta, \varphi)$,

2.1 The tokamap

depending on the poloidal and toroidal angles,

$$H = H_0(\psi) + \varepsilon H_1(\psi, \theta, \varphi) , \qquad H_0(\psi) = \int \frac{d\psi}{q(\psi)} . \qquad (2.1.3)$$

Here the dimensionless perturbation parameter ε stands for the relative strength of the magnetic perturbations. The perturbation Hamiltonian H_1 is a 2π-periodic function of θ and φ, which can be always presented as a Fourier series

$$H_1(\psi, \theta, \varphi) = \sum_{m,n} H_{mn}(\psi) \cos(m\theta - n\varphi + \chi_{mn}) . \qquad (2.1.4)$$

The integers m and n are called the poloidal and toroidal mode numbers, respectively. The constants χ_{mn} represent their phases.

The most powerful tool to study magnetic field lines is the Poincaré map, which replaces the continuous system (2.1.2) by the discrete one

$$(\psi_{k+1}, \theta_{k+1}) = \hat{P}(\psi_k, \theta_k) . \qquad (2.1.5)$$

It relates the k−th intersection point, (ψ_k, θ_k), of the field line with the poloidal section $\varphi =$ constant to the next one $(\psi_{k+1}, \theta_{k+1})$ after one toroidal turn, i.e. it is a return map of variables (ψ, θ) to a certain poloidal section $\varphi =$ constant. Such a map should be flux-preserving, i.e. $|\partial(\psi_{k+1}, \theta_{k+1})/\partial(\psi_k, \theta_k)| = 1$.

To avoid time-consuming small-step integrations of field line equations, different analytical mapping models for the Poincaré map (2.1.5) have been proposed (see [8] for references). Such a map, the tokamap, being compatible with the toroidal geometry, has been proposed by Balescu et al. [11]. It is of the following form

$$\psi_{k+1} = \psi_k - \varepsilon \frac{\psi_{k+1}}{1 + \psi_{k+1}} \sin(\theta_k) , \qquad (2.1.6)$$

$$\theta_{k+1} = \theta_k + \frac{2\pi}{q(\psi_{k+1})} - \varepsilon \frac{1}{(1 + \psi_{k+1})^2} \cos(\theta_k) . \qquad (2.1.7)$$

It should be emphasized that it has not been rigorously derived from a continuous Hamiltonian system such as (2.1.2), (2.1.3), and (2.1.4).

Here we gave a short outline of the tokamap model by Balescu. In general, the postulated model is not an appropriate ansatz to analyze chaotic systems systematically. We need to derive a mapping procedure from the physically motivated Hamiltonian system like (2.3.1) in a more systematic way. For this, we will use the symmetric mapping technique by S. Abdullaev et al.

2.2 The mapping technique

The mapping technique is an advanced procedure to calculate the so called Poincaré plot of the system. The Poincaré plot is a two dimensional symplectic map of the action-angle variables (ψ, θ), where each point in the map is an intersection point of a system's trajectory with the (ψ, θ) plane at discrete times t_k, $(k = 0, \pm 1, \pm 2, \ldots)$, while all other variables are fixed. Usually the Poincaré map is calculated by the numerical integration of the Hamiltonian equations of motion, which are a system of first order ordinary differential equations. Then the (ψ, θ) values at the discrete times t_k are plotted. Using the mapping technique, these numerical long time calculations are reduced to algebraic iteration equations, where one iteration step corresponds to the integration of the equations of motion across one time step $k \to k+1$. Symplectic maps have been used in physics for a long time, but the mapping technique in its general symmetric form, which has been developed by S. Abdullaev et al. [4, 5, 9, 10], is a relatively new technique to construct a symplectic symmetric map.

The mapping technique is based on Hamiltonian systems, described by the Hamiltonian $H(p, q, t)$ with the canonical momentum p, the coordinate q and the time t. For the reason of simplicity, we focus on a one dimensional system, which also depends explicitly on time, a system with so called $1\frac{1}{2}$ degrees of freedom. Such a system is the system with the least number of degrees of freedom, which can show deterministic chaotic behavior. One application of such a system is, e.g., the Hamiltonian description of magnetic field lines, which dynamics are of great interest. They are given by the Hamiltonian equations of motion

$$\frac{dp}{dt} = \frac{\partial H}{\partial q} \quad \text{and} \quad \frac{dq}{dt} = -\frac{\partial H}{\partial p} \ . \tag{2.2.1}$$

In the following we assume that the system is described by an integrable Hamiltonian H_0, perturbed by a small perturbation εH_1 with $\varepsilon \ll 1$. Then the total Hamiltonian reads

$$H(p, q, t) = H_0(p, q) + \varepsilon H_1(p, q, t) \ . \tag{2.2.2}$$

Using the Hamilton-Jacobi theory, see e.g. [26], on the integrable part H_0, we can transform the variables (p, q) globally to action-angle variables $(p, q) \to (\psi, \theta)$ by canonical transformations. The transformed Hamiltonian then only depends on the action variable ψ, meaning $H_0 = H_0(\psi)$. Now we can solve the Hamiltonian equations of motion very easily and conclude that in this case ψ is a constant of motion and

$$\theta = \theta_0 + \frac{\partial H_0}{\partial \psi}(t - t_0) \tag{2.2.3}$$

is the unperturbed trajectory with $\theta_0 = \theta(t_0)$.

2.2 The mapping technique

The perturbed system
$$H(\psi, \theta, t) = H_0(\psi) + \varepsilon H_1(\psi, \theta, t) \tag{2.2.4}$$
is usually a non-integrable system so that a global transformation to action-angle variables can no longer be applied. Nevertheless, we can use canonical transformations to reduce the description of the dynamics of the perturbed system from the differential equations of motion (2.2.1) to a system of algebraic equations, the Hamiltonian map. This procedure is called the mapping technique. The map
$$(\theta_{k+1}, \psi_{k+1}) = M(\theta_k, \psi_k) \tag{2.2.5}$$
relates the variables $\theta_k = \theta(t_k)$ and $\psi_k = \psi(t_k)$, given at the time t_k, to the variables $\theta_{k+1} = \theta(t_{k+1})$ and $\psi_{k+1} = \psi(t_{k+1})$ at the time t_{k+1}, with $t_k = k\tau$, ($k = 0, \pm 1, \pm 2, \ldots$), where τ is a finite time interval. Within this time interval, which means $t_k < t < t_{k+1}$, we can transform the variables (ψ, θ) to new variables (ξ, ϑ) in such a way that ξ is again a constant of motion, but now only within this time interval. The generating function $F(\xi, \theta, t)$ of this transformation then fulfills the Hamilton-Jacobi equation
$$H_0\left(\frac{\partial F}{\partial \theta}\right) + \varepsilon H_1\left(\theta, \frac{\partial F}{\partial \theta}, t\right) + \frac{\partial F}{\partial t} = \hat{H}(\xi) \tag{2.2.6}$$
within this time interval, where $\hat{H}(\xi)$ is the new transformed Hamiltonian. Because the Hamiltonian system consists of a main part and a small perturbation, proportional to ε, we can expand the generating function F into a power series of ε
$$F(\xi, \theta, t) = F_0(\xi, \theta, t) + \varepsilon F_1(\xi, \theta, t) + \ldots, \tag{2.2.7}$$
where F_0 describes the transformation of the unperturbed system. From the Hamilton-Jacobi equation we obtain
$$F_0 = \theta\xi - H_0(\xi)t + \hat{H}(\xi, t)t, \tag{2.2.8}$$
with the relations between the old and new variables
$$\psi = \frac{\partial F_0}{\partial \theta} = \xi, \quad \vartheta = \frac{\partial F_0}{\partial \xi} = \theta - \frac{\partial H_0}{\partial \xi}t + \frac{\partial \hat{H}}{\partial \xi}t. \tag{2.2.9}$$
This transformation is a trivial one, because the unperturbed part is already transformed to action-angle variables, so that the resulting map
$$\psi_{k+1} = \psi_k, \quad \theta_{k+1} = \theta_k + \frac{\partial H_0}{\partial \psi_{k+1}}(t_{k+1} - t_k) \tag{2.2.10}$$
directly corresponds to the solution of the equations of motion (2.2.3).

Chapter 2. Formation of chaos in the symmetric tokamap regime

Now we take the next order in ε of the generating function into account and we define $F_1(\xi, \theta, t) =: S(\xi, \theta, t)$, calling now S the generating function. Then we can construct the map in the following way. At the beginning of the time interval t_k we transform to the new variables (ξ, ϑ), pass the time interval according to Eq. (2.2.10) and at the end of the interval t_{k+1} we transform back to the old variables (ψ, θ), but now at the time t_{k+1}. Therefore, we can derive the map

$$\xi_k = \psi_k - \varepsilon \frac{\partial S_k}{\partial \theta_k}, \qquad \vartheta_k = \theta_k + \varepsilon \frac{\partial S_k}{\partial \xi_k} \qquad (2.2.11)$$

$$\vartheta_{k+1} = \vartheta_k + \frac{\partial H_0(\xi_k)}{\partial \xi_k}(t_{k+1} - t_k) \qquad (2.2.12)$$

$$\psi_{k+1} = \xi_k + \varepsilon \frac{\partial S_{k+1}}{\partial \theta_{k+1}}, \qquad \theta_{k+1} = \vartheta_{k+1} - \varepsilon \frac{\partial S_{k+1}}{\partial \xi_k} \qquad (2.2.13)$$

with $S_k = S(\xi_k, \theta_k, t_k)$ and $S_{k+1} = S(\xi_k, \theta_{k+1}, t_{k+1})$. This is the common form of the iteration procedure the mapping technique is based on, and all maps presented in this thesis are related to this form. Note, the first and the last equation are implicit ones, which usually have to be solved numerically by using for example Newton's method.

The most important part for the mapping technique is the determination of the generating function S. Therefore, we expand Eq. (2.2.6) with respect to ε and obtain for the first order

$$\frac{\partial S}{\partial t} + \frac{\partial H_0}{\partial \xi}\frac{\partial S}{\partial \theta} = \hat{H}_1(\xi) - H_1(\xi, \theta, t) . \qquad (2.2.14)$$

The left side of this equation is equal to the total time derivative of S along the unperturbed trajectory $\theta(t) = \theta_0 + \frac{\partial H_0}{\partial \xi}(t - t_0)$, given by the unperturbed Hamiltonian H_0. So we can integrate the equation and get

$$S(\xi, \theta, t) = \hat{H}_1(\xi)(t - t_0) - \int_{t_0}^{t} H_1(\xi, \theta(t'), t') \, dt' \qquad (2.2.15)$$

with the free parameters t_0 and $\hat{H}_1(\xi)$, which can be chosen arbitrarily, but only some special choices are convenient. The integration can be performed, if we assume a special Fourier ansatz

$$H_1(\xi, \theta, t) = H_0^{(1)}(\xi) + \sum_{m,n} H_{mn}(\xi) e^{im\theta - in\omega t} \qquad (2.2.16)$$

for the perturbation part H_1 of the Hamiltonian system. With $H_0^{(1)}(\xi) = -\hat{H}_1(\xi)$, which is the most convenient choice for $\hat{H}_1$, we find

$$S(\xi,\theta,t) = -(t-t_0)\sum_{m,n}[a(x_{mn}) - ib(x_{mn})]H_{mn}(\xi)e^{im\theta-in\omega t} \qquad (2.2.17)$$

with $x_{mn} = (t-t_0)[m\frac{\partial H_0}{\partial \xi} - n\omega]$, $a(x) = \frac{\sin x}{x}$ and $b(x) = \frac{1-\cos x}{x}$.

The parameter t_0, which is a reference time within the finite time interval τ, is still arbitrary. The proper choice of t_0 decides, which kind of map is created. Choosing t_0 exactly in the middle of the time interval

$$t_0 = \frac{1}{2}(t_{k+1} + t_k) \,, \qquad (2.2.18)$$

we get the map in its symmetric form. This type of map is called symmetric map, because of its invariance to inversion of time. This is the mapping procedure, which will be used in this thesis, because it is in very good agreement with the direct numerical integration of the equations of motion [9]. There are also nonsymmetric maps, so called twist maps [4, 9], which are created, if one chooses t_0 to be the beginning t_k or end t_{k+1} of the time interval. But, as one can see from the references, the twist maps are of less accuracy and do not conserve all symmetries, especially the time inversion, of the original Hamiltonian system.

Here we have only shown the derivation of the first order generating function. One can also calculate all higher orders of F successively from the expansion of Eq. (2.2.6). Including higher orders of the generating function would improve the accuracy, but, due to the smallness of the perturbation, these corrections are negligible small. If the accuracy is not sufficient, one can reduce the time step size τ. Such a reduction also improves the convergence of Newton's method, when used on the implicit equations.

2.3 The tokamap in its symmetric form

Now we will present the so called symmetric tokamap, which is a modification of the original tokamap model. In contrast to the originally proposed non-symmetric tokamap [11], the symmetric tokamap is based on a continuous Hamiltonian system, consisting of an integrable part, specified by the safety factor, and a non-integrable perturbation. The model is close to the kicked rotator, with a nonlinear dependency of the amplitude on the flux ψ. The Hamiltonian system has a fundamental symmetry according to the transformation $t \to -t$ and $H \to -H$. This time-reversal symmetry is not reflected in the usual (non-symmetric) form of the original tokamap. Due to this lack of symmetry in the non-symmetric tokamap, the symmetric tokamap was proposed [8]. The symmetric tokamap can be easily inverted in time and the inverse map has the same structure and Poincaré section. The symmetric tokamap is in very good agreement with the Poincaré section of the continuous system [8].

Chapter 2. Formation of chaos in the symmetric tokamap regime

It was supposed [27] that the tokamap corresponds to the Hamiltonian

$$H = H_0(\psi) + \varepsilon H_1(\psi, \theta, \varphi) = \int \frac{d\psi}{q(\psi)} + \varepsilon \frac{\psi}{1+\psi} \cos(\theta) \sum_{k=-\infty}^{\infty} \delta(\varphi - 2\pi k) . \quad (2.3.1)$$

A derivation of the tokamap from the Hamiltonian (2.3.1) encounters a difficulty, related with the presence of delta functions (see [8] and references therein). In order to avoid these difficulties, the following regularization procedure has been proposed in Ref. [8]. Let us consider the Hamiltonian

$$H = H_0(\psi) + \varepsilon H_1(\psi, \theta, \varphi) = \int \frac{d\psi}{q(\psi)} + \varepsilon \frac{\psi}{1+\psi} \cos(\theta) \sum_{s=-M}^{M} \cos(s\varphi) , \quad (2.3.2)$$

containing the sum of a finite number M of trigonometric functions. Using the Poisson summation rule, it is easy to see that the Hamiltonian (2.3.1) follows from the regularized Hamiltonian (2.3.2) in the limit $M \to \infty$.

Applying the construction of canonical mappings, presented in Sec. 2.2, to the Hamiltonian (2.3.2), and performing the limit $M \to \infty$, one obtains the generating function

$$S = -\varepsilon \frac{\psi}{1+\psi} \cos(\theta) , \quad (2.3.3)$$

which leads to the exact mapping [8]

$$\xi_k = \psi_k - \frac{\varepsilon}{2} \frac{\xi_k}{1+\xi_k} \sin(\theta_k) , \qquad \vartheta_k = \theta_k - \frac{\varepsilon}{2} \frac{1}{(1+\xi_k)^2} \cos(\theta_k) , \quad (2.3.4)$$

$$\vartheta_{k+1} = \vartheta_k + \frac{2\pi}{q(\xi_k)} , \quad (2.3.5)$$

$$\psi_{k+1} = \xi_k - \frac{\varepsilon}{2} \frac{\xi_k}{1+\xi_k} \sin(\theta_{k+1}) , \qquad \theta_{k+1} = \vartheta_{k+1} - \frac{\varepsilon}{2} \frac{1}{(1+\xi_k)^2} \cos(\theta_{k+1}) . \quad (2.3.6)$$

This map is called the symmetric tokamap. Note, the non-symmetric tokamap (2.1.6) cannot be derived from the Hamiltonian (2.3.2), using the regularization method. Generally, the symmetric tokamap is an implicit map, but the first equation (2.3.4) can be explicitly resolved with respect to ξ_k,

$$\xi_k = \frac{1}{2} \left[\sqrt{P^2(\psi_k, \theta_k) + 4\psi_k} - P(\psi_k, \theta_k) \right] , \quad (2.3.7)$$

where

$$P(\psi_k, \theta_k) = 1 - \psi_k + \frac{\varepsilon}{2} \sin(\theta_k) . \quad (2.3.8)$$

The ψ-equation of the non-symmetric tokamap (2.1.6) can be resolved in the same way. The implicit equation for θ_{k+1} cannot be solved analytically. For this purpose we use the Newton

16

method.

The symmetric tokamap is invariant with respect to the transformation $k \leftrightarrow k+1$ when $\varepsilon \to -\varepsilon$ and $q \to -q$. So it can easily be inverted. This property corresponds to the symmetry of the continuous Hamiltonian system with respect to the transformations $t \to -t$ and $H \to -H$. Typical plots of the symmetric and non-symmetric tokamap are shown in the Figs. 2.1 and 2.2, respectively, for the same perturbation parameter $\varepsilon = 4.5/2\pi$ and a monotonic q-profile.

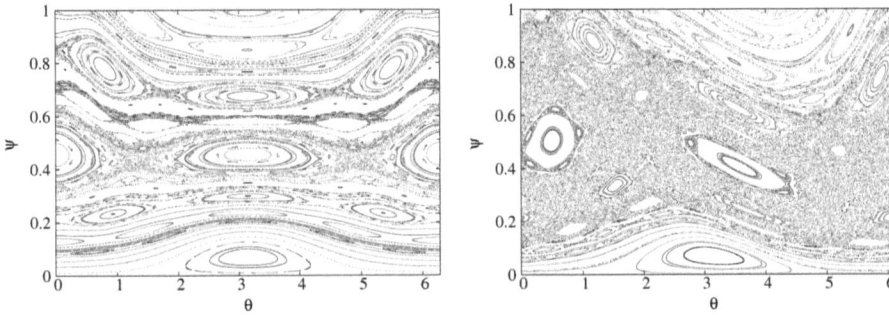

Figure 2.1: Poincaré plot of the symmetric tokamap with a monotonic q-profile and $\varepsilon = 4.5/2\pi$.

Figure 2.2: Same as Fig. 2.1, but now for the non-symmetric tokamap.

We are using

$$q(\psi) = \frac{4}{(2-\psi)(2-2\psi+\psi^2)} \quad (2.3.9)$$

for the monotonic q-profile. Later we shall generalize to a non-monotonic q-profile of the form

$$q(\psi) = \frac{q_m}{1 - a(\psi - \psi_m)^2}, \quad (2.3.10)$$

corresponding to a reverse magnetic shear profile in a tokamak. q_m is the minimum value of q at $\psi = \psi_m$. The parameters $\psi_m = \left[1 + \left(\frac{1-q_m/q_1}{1-q_m/q_0}\right)^{1/2}\right]^{-1}$ and $a = \frac{1-q_m/q_0}{\psi_m^2}$ are evaluated with the values $q_0 = q(0)$ and $q_1 = q(1)$. We will choose $q_0 = 3$, $q_1 = 6$, and $q_m = 1.5$, so $\psi_m \approx 0.5505$. Further we introduce the winding number $\Omega(\psi) = 1/q(\psi)$, given by the reciprocal value of the safety factor.

2.4 Statistical properties of the tokamap

As it can be seen from the Figs. 2.1 and 2.2, there are some fundamental differences between both forms of the tokamap. The symmetric one is less chaotic than the non-symmetric one, which will become quantitatively more evident when we analyze the break up of KAM surfaces. The periodic points are located at different positions. Only the fixed points, defined by $\psi_{k+1} =$

Chapter 2. Formation of chaos in the symmetric tokamap regime

ψ_k and $\theta_{k+1} = \theta_k$, which are also periodic points with period 1, are the same in both maps. There is one elliptic point on the equatorial plane at $\theta = \pi$, and one hyperbolic point in the center at $\psi = 0$ (formally at $\theta = \pi/2$ and $\theta = 3\pi/2$). The periodic points, defined by $\psi_{k+n} = \psi_k$ and $\theta_{k+n} = \theta_k$ for $n \geq 2$, are different. In the symmetric tokamap the periodic points are symmetrically arranged to $\theta = \pi$, and for all island-chains there is one elliptic point at $\theta = \pi$. This fact is related to the invariance with respect to the translation $\theta \leftrightarrow \pi - \theta$ of the symmetric tokamap, which results from the time invariance. The non-symmetric tokamap does not show such a symmetry. The following analysis of the statistical properties will detect more qualitative and quantitative differences, but also common properties. It will help us to classify the chaotic system as such.

At first we analyze the transport (diffusion) of magnetic field lines, especially for large perturbations. Therefore, we calculate the mean square displacement (MSD)

$$\sigma(t) = \langle (\psi(t) - \langle \psi(t) \rangle)^2 \rangle \qquad (2.4.1)$$

of the flux, where $\langle \ldots \rangle$ stands for an averaging over initial points. Here we do not terminate the ψ-values at $\psi = 1$; in later sections we shall discuss the footprints when a wall is set at $\psi = 1$. The numerical calculations show that the MSD increases linearly in time for large perturbations $\varepsilon \gg 1$, meaning normal diffusion of the flux ψ. The running diffusion coefficient

$$D = \frac{1}{2}\frac{d\sigma(t)}{dt} = \text{const} \qquad (2.4.2)$$

still depends on the perturbation parameter ε. The evaluations are similar to those for the standard map

$$\psi_{k+1} = \psi_k + K \sin(\theta_k), \qquad (2.4.3)$$
$$\theta_{k+1} = \theta_k + \psi_{k+1}. \qquad (2.4.4)$$

In Fig. 2.3 the diffusion coefficient, normalized with the quasi-linear diffusion coefficient $D_{ql} \equiv \frac{\varepsilon^2}{4}$ of the standard map, is plotted versus the perturbation parameter ε for very large values of the latter. The solid line shows the tokamap result. As one can see, the diffusion coefficient is proportional to ε^2, but the factor is not equal to $1/4$ as it is for the standard map (dashed line). The quasi-linear diffusion coefficient for the tokamap is given by

$$D \approx 0.36 \cdot \frac{\varepsilon^2}{4} = 0.36 \cdot D_{ql}. \qquad (2.4.5)$$

For the standard map we recognize the typical oscillations around D_{ql}, already known in literature [28, 23]. For the tokamap such regular oscillations are not observed. The reason is

2.4 Statistical properties of the tokamap

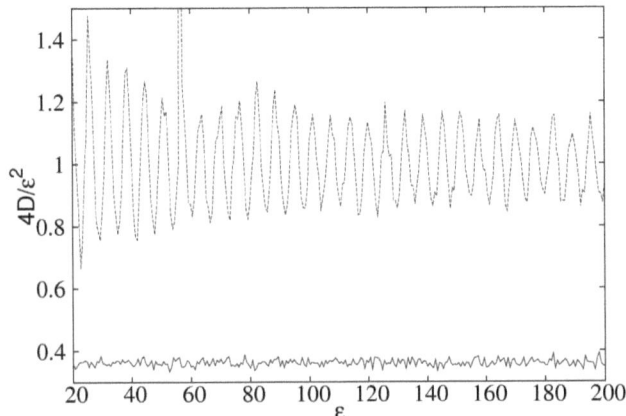

Figure 2.3: Diffusion coefficients for the symmetric tokamap (solid line) and the standard map (dashed line), normalized with $\varepsilon^2/4$. The diffusion coefficient is plotted versus the perturbation parameter ε.

the nonlinear winding number $\Omega(\psi) \sim \psi^3$. Using the standard map with such a non-linear winding number, the diffusion coefficient of the standard map also shows irregular fluctuations and no regular oscillations around D_{ql}. The factor 0.36 reflects the non-even distribution of the ψ-values for large perturbations. Although the mean of ψ goes to infinity, when time goes to infinity, the ψ-values are extremely dense at $\psi = 0$.

The quasi-linear diffusion coefficient for the tokamap is valid in the regions of large perturbations, but also for moderate perturbations $\varepsilon \sim 1$ and large fluxes $\psi > 3$. Both forms of the tokamap show the same quasi-linear behavior.

For small perturbations, $\varepsilon < 0.7$, both forms of the tokamap have (two last) intact KAM surfaces. The lower one (near the core) separates the period-1 island core from the chaotic sea around the period-2 island chain. The second (upper one) separates the chaotic sea from the outside area, which roughly begins at $\psi \approx 1$.

For applications, the exact values of the critical perturbations for destruction of the KAM surfaces are highly relevant. Here we detect a major difference between the symmetric and the non-symmetric tokamap. The critical perturbations are:

- Symmetric tokamap:
  upper KAM surface: $\varepsilon = 5.414/2\pi$,   lower KAM surface: $\varepsilon = 5.719/2\pi$.

- Non-symmetric tokamap:
  upper KAM surface: $\varepsilon = 4.998/2\pi$,   lower KAM surface: $\varepsilon = 4.857/2\pi$.

Chapter 2. Formation of chaos in the symmetric tokamap regime

On the one hand, the critical perturbations are definitely higher for the symmetric tokamap than for the non-symmetric one. But more important is that the KAM surfaces are breaking in a different order. For the symmetric tokamap the lower KAM surface near the core is destroyed after the upper one. This means that, while the field lines within the chaotic sea can reach the outside area and are getting lost, the core area is still protected. For the non-symmetric map it is the other way round. The field lines near the core can diffuse into the chaotic sea, while the whole chaotic area $\psi < 1$ is still confined. So there is not only the quantitative difference that the symmetric tokamap is more robust against perturbations. There is also the qualitative difference in the order of the break-up of KAM surfaces. Having in mind that the symmetric tokamap has a direct link to the continuous Hamiltonian system, resulting from physically motivated perturbations, the results from the symmetric tokamap are more useful for quantitative predictions.

For perturbations larger than the critical one, the transport behavior of the symmetric tokamap through the open KAM surface is similar to that of the non-symmetric map. Although the surface is open, close to the critical perturbation value, the mean square displacement (MSD) does not change for a long time, up to 10^6 iterations. After that time a sudden and significant increase of the MSD can be observed. This has already been analyzed in [29] for the non-symmetric tokamap. The same behavior can be observed for the symmetric tokamap, as it is shown in the Figs. 2.4 and 2.5. There, the MSD is plotted on a log-time scale for the perturbation parameters $\varepsilon = 4.875/2\pi$ and $\varepsilon = 6/2\pi$, respectively. In Fig. 2.4 we can see that for the symmetric tokamap the lower KAM surface is still intact.

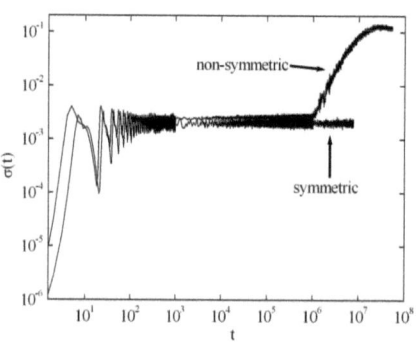

Figure 2.4: MSD of the flux, plotted versus time (iteration number) in double log-scale, with $\varepsilon = 4.875/2\pi$ for the symmetric and the non-symmetric tokamap. We have chosen 1000 starting points at $\psi = 0.001$ and $0 \leq \theta \leq 2\pi$.

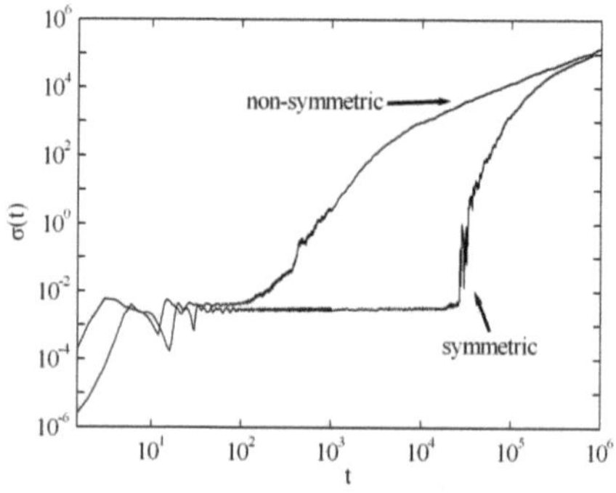

Figure 2.5: Same as Fig. 2.4, but for $\varepsilon = 6/2\pi$.

In the Figs. 2.4 and 2.5 we recognize that the characteristic time T when the MSD starts increasing depends on the perturbation parameter, especially on the difference $\varepsilon - \varepsilon_c$, where ε_c is the critical perturbation for the break-up of the KAM surface. The number N of "test-particles" (i.e. fictitious particles sticking to the field lines) staying below the already broken KAM surface decreases exponentially in time, when $\varepsilon > \varepsilon_c$. We have

$$\frac{N}{N_0} = \begin{cases} e^{-\lambda t}, & \varepsilon \geq \varepsilon_c, \\ 1, & \varepsilon < \varepsilon_c. \end{cases} \qquad (2.4.6)$$

The scaling is

$$\lambda \sim (\varepsilon - \varepsilon_c)^3 \quad \Rightarrow \quad T \sim (\varepsilon - \varepsilon_c)^{-3} \qquad (2.4.7)$$

for the dependence on the perturbation parameter. This result is valid for both KAM surfaces and both maps.

2.5 Construction of stable und unstable manifolds

As mentioned above, we will use the concept of the stable and unstable manifolds of unstable objects, hyperbolic periodic points in our case, to obtain more information about the formation of chaotic layers and the transport within. Although the concept of the stable and unstable manifolds is already known [12], it has never been applied to the tokamap model before and as we shall see, the analysis of the structures and the dynamics of the manifolds will lead to interesting new results.

The unstable manifold of a hyperbolic periodic point is defined as the set of points which converge under the map towards the periodic point for $n \to -\infty$. The stable manifold is the unstable manifold of the inverse map. Each manifold has two opposite sides, one on the left of the hyperbolic point, the other on the right of it. During the calculation of the manifold one has to watch that one does not change the sides. When n is the period of the hyperbolic point, then $F(\vec{x})$ stands for the $2n$-th iteration of $\vec{x}$. There are some examples, where the right side of the manifold is projected to the left side after n iterations. Using $2n$ iterations, the side of the manifold will be preserved.

To calculate the unstable manifold of a hyperbolic point with period n, we choose a starting-point $\vec{x}$ very close to the hyperbolic point. Then all iterations of this point are lying on the unstable manifold or extremely close to it. This can be understood by the way, the points are following the manifolds. We can split the directional vector whom the point $\vec{x}$ follows during $2n$ iterations in one part along the stable manifold and one part along the unstable manifold. Due to its unstable component, the point will follow the unstable manifold away from the hyperbolic point. Due to its stable component, the point will be driven closer to the unstable manifold. Therefore, the small error that the starting point is not exactly located

Chapter 2. Formation of chaos in the symmetric tokamap regime

on the unstable manifold, will diminish rapidly during the iteration. But the error has to be very small, meaning that the starting point must be located very close to the hyperbolic point, which makes it absolutely necessary to know the position of the hyperbolic point precisely, see Sec. 2.6.

When we have chosen the starting point $\vec{x}$, we calculate the $2n$-th iterate $\vec{y} = F(\vec{x})$ of $\vec{x}$. The line-element between $\vec{x}$ and $\vec{y}$ then approximates the unstable manifold very well, because the unstable manifold is not curved that close to the hyperbolic point and as already mentioned, the point $\vec{y}$ is located on or even closer to the unstable manifold than $\vec{x}$. By iterating the line-element in the following way [12], we get the unstable manifold.

When $G_k(\vec{x}) = F^k(\vec{x}) = F(F(\ldots F(\vec{x})\ldots))$, we start with $\vec{x}_0 = \vec{x}$ and $k = 1$. Then we choose points on the line-element $\vec{x}_i$ in such a way that the distance between $G_k(\vec{x}_{i-1})$ and $G_k(\vec{x}_i)$ is less than a chosen maximum distance. Is the distance between $G_k(\vec{x}_{i-1})$ and $G_k(\vec{x}_i)$ less than a chosen minimum distance, one can use a slightly larger step on the line-element for the choice of x_{i+1}. The transition from k to $k+1$ is automatically given by $G_k(\vec{y}) = G_{k+1}(\vec{x})$. As one can see, the step size on the line-element regulates itself so that the distance of two neighboring points of the manifold plot is always less than the maximum distance, but mostly larger than the minimum distance. One has to consider that with increasing k the step size on the line-element will decrease, until it reaches the numerical accuracy of the machine. Then the manifold can no longer be traced, but this point highly depends on the choice of the maximum and minimum distance.

Using this method, one side of the unstable manifold can be calculated very precisely, because numerical errors cancel themselves out, as already mentioned above. The other side of the unstable manifold can be calculated by choosing the starting point $\vec{x}$ on the other side of the hyperbolic point for example. The stable manifold is the unstable manifold of the inverse map and can be calculated by using the inverse iteration.

2.6 Periodic points

To calculate the stable and unstable manifolds, one has to know the positions of the periodic points very precisely. Here we outline an algorithm to determine the periodic points of a map.

A periodic point with period n is defined through

$$\psi = M_\psi^n(\psi, \theta) \qquad \theta = M_\theta^n(\psi, \theta) \bmod 2\pi , \qquad (2.6.1)$$

with M_ψ^n and M_θ^n being the n-times iterations of the map with respect to ψ and θ, respectively. There are two different kinds of periodic points. The elliptic ones, which are at the centers of the islands, are stable. A trajectory very close to the elliptic point will iterate on an elliptic orbit around that periodic point and will always stay close to it. These elliptic periodic points are

2.6 Periodic points

only of low interest to us. The hyperbolic ones, i.e. the intersection points of the unperturbed separatrix, are located between the islands. They are unstable. A trajectory close to the hyperbolic point will follow a hyperbolic orbit away from the periodic point. These points and their stable and unstable manifolds are the source of chaos and anomalous transport, as we shall see later.

Finding the hyperbolic points is extremely difficult, due to their unstable character. But they can be determined numerically, using a minimization method [29, 30]. The problem is similar to solving a system of N nonlinear equations

$$F_i(\vec{x}) = F_i(x_1, x_2, \ldots, x_N) = 0 \quad 1 \leq i \leq N , \qquad (2.6.2)$$

while we only consider $N = 2$ here. We have to minimize

$$f(\vec{x}) = \sum_{i=1}^{N} (F_i(\vec{x}))^2 . \qquad (2.6.3)$$

Therefore, we consider the function

$$g(t) = f(\vec{a} + t \cdot \vec{d}) , \qquad (2.6.4)$$

where $\vec{a} = (a_1, \ldots, a_N)$ is a chosen starting point and $\vec{d} = (d_1, \ldots, d_N)$ is a chosen directional vector. Now we have to find the minimum t_{min} of $g(t)$. The next starting point is then given by $\vec{a}_1 = \vec{a} + t_{min} \cdot \vec{d}$, and we have to choose a new direction $\vec{d}_1$ in which the one-dimensional minimization is then performed. To ensure the convergence, the proper choice of the direction is important. The native choice for the direction would be the gradient

$$\vec{d} = -\nabla f(\vec{a}) , \qquad (2.6.5)$$

but the conjugated direction method [29], which is outlined in the following, is more effective.

For one N-dimensional minimization step, we use N sub-steps

- Sub-step 0:
  For the starting point $\vec{x}_0$, we choose the direction

  $$\vec{d}_0 = -\nabla f(\vec{x}_0) .$$

  By minimizing the function

  $$g_0(t) = f(\vec{x}_0 + t \cdot \vec{d}_0)$$

Chapter 2. Formation of chaos in the symmetric tokamap regime

with respect to t, we get the new starting point

$$\vec{x}_1 = \vec{x}_0 + t_{min} \cdot \vec{d}_0 .$$

- Sub-step $k+1$ $(k < N-1)$:
  Using the directional vector

$$\vec{d}_{k+1} = -\nabla f(\vec{x}_{k+1}) + \beta_k \vec{d}_k$$

with

$$\beta_k = \frac{||\nabla f(\vec{x}_{k+1})||^2}{||\nabla f(\vec{x}_k)||^2} ,$$

we minimize $g_{k+1}(t) = f(\vec{x}_{k+1} + t \cdot \vec{d}_{k+1})$ with respect to t and calculate $\vec{x}_{k+2} = \vec{x}_{k+1} + t_{min} \cdot \vec{d}_{k+1}$.

These steps are performed up to $N-1$, then the procedure starts again at sub-step 0 with $\vec{x}_0 = \vec{x}_N$.

The one-dimensional minimization is performed in each sub-step, using Newton's method to find the zero point of the first derivative $g'(t)$. Usually the derivative of the considered function, which is the second one $g''(t)$ here, is needed for Newton's method. With $\vec{y} = \vec{x} + t \cdot \vec{d}$ we obtain

$$g'(t) = \nabla f(\vec{y}) \cdot \vec{d} = 2 \sum_{i,j=1}^{N} \frac{\partial F_j}{\partial x_i}(\vec{y}) F_j(\vec{y}) d_i \qquad (2.6.6)$$

and

$$g''(t) = 2 \sum_{i,j,l=1}^{N} \frac{\partial^2 F_j}{\partial x_i \partial x_l}(\vec{y}) F_j(\vec{y}) d_i d_l + 2 \sum_{i,j,l=1}^{N} \frac{\partial F_j}{\partial x_i} \frac{\partial F_j}{\partial x_l}(\vec{y}) d_i d_l . \qquad (2.6.7)$$

One can neglect the first term on the right side of the second derivative so that the second derivative can be approximated by products of the first derivatives. Note, all values are those of the actual sub-step.

Using this simplification, we can approximate t_{min} by

$$t_{min} = -\frac{g'(0)}{\tilde{g}''(0)} , \qquad (2.6.8)$$

with

$$\tilde{g}''(t) = 2 \sum_{i,j,l=1}^{N} \frac{\partial F_j}{\partial x_i} \frac{\partial F_j}{\partial x_l}(\vec{y}) d_i d_l . \qquad (2.6.9)$$

This procedure converges very fast and will lead to both types of periodic points up to the desired accuracy. The periodic point being found, depends on the choice of the first starting point. The best way to find periodic points in a certain area, is to use a grid of starting points.

This is only one possible algorithm to find periodic points. Another very good one is the two dimensional Newton method, which will not be described here.

2.7 Stable and unstable manifolds of the symmetric tokamap

The statistical properties of the tokamap, which are shown in Sec. 2.4, are the visible and measurable results of the chaotic dynamics, but they cannot explain, why the chaotic layers are formed and how the transport is generated. The analysis of the influence of the hyperbolic periodic points and their stable and unstable manifolds on chaotic motion, appearing of chaos and chaotic transport is the main task.

We will show that the hyperbolic points are the source of chaos. Due to the slightest perturbation, the ideal unperturbed separatrix splits into the stable and unstable manifolds.

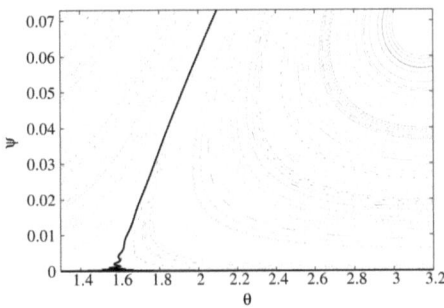
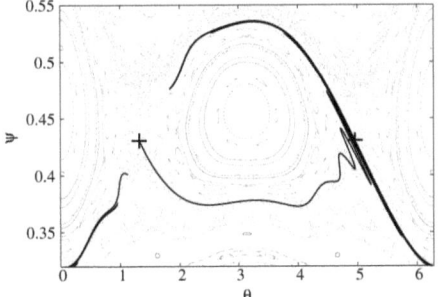

Figure 2.6: Homoclinic unstable manifold (solid line) of the period-1 hyperbolic point at $\psi = 0$ and $\theta = \pi/2$ of the symmetric tokamap, with monotonic q-profile and $\varepsilon = 4.5/2\pi$.

Figure 2.7: Heteroclinic right-hand-sided unstable manifold of the period-2 hyperbolic point at $\psi = 0.431$ and $\theta = 1.324$ of the symmetric tokamap with monotonic q-profile and $\varepsilon = 4.5/2\pi$. The hyperbolic points are marked by crosses.

Figure 2.6 shows the unstable manifold (solid line) of the period-1 hyperbolic point at $\psi = 0$ and $\theta = \pi/2$ of the symmetric tokamap, with monotonic q-profile and $\varepsilon = 4.5/2\pi$, plotted very close to the periodic point, and Fig. 2.7 shows the right-hand-sided unstable manifold (solid line) of the period-2 hyperbolic point at $\psi = 0.431$ and $\theta = 1.324$. The manifolds are plotted only up to a certain length to ensure the visibility of their behavior. The total manifold fills the entire chaotic area and has an infinite length.

As one can see, the unstable manifold starts oscillating, when it approaches another hyperbolic point (heteroclinic) or the same (homoclinic), and the amplitude increases strongly. When the amplitude becomes large enough so that the loops also come close to a hyperbolic point, the loops start oscillating too. In this way the unstable manifold becomes dense in the

Chapter 2. Formation of chaos in the symmetric tokamap regime

chaotic area for a closed chaotic system. The stable manifolds show the same oscillatory behavior close to hyperbolic points. Important is the fact that unstable manifolds never intersect with each other, but the stable manifolds intersect with the unstable ones. Their oscillations intersect infinite times, while the area enclosed by the intersection is preserved. Field lines within such enclosed areas are iterating from one area to another [13]. These areas are getting very long and extremely thin. Due to their stable components, they are iterating towards the hyperbolic point, while due to their unstable components, they are elongated in the directions of the unstable manifold. Close to hyperbolic points (note the increasing amplitude of the oscillations and the area preserving property) two neighboring field lines are always located in such different areas and therefore iterate in completely different ways. Assuming that two field lines are located in the same area, this area iterates towards the hyperbolic point, as described above, and comes automatically into regions of a higher rate of intersection. The structure formed by the intersections of stable and unstable manifolds is a fractal one. This means that similar initial conditions always leads to totally different results. Chaos appears around the hyperbolic points.

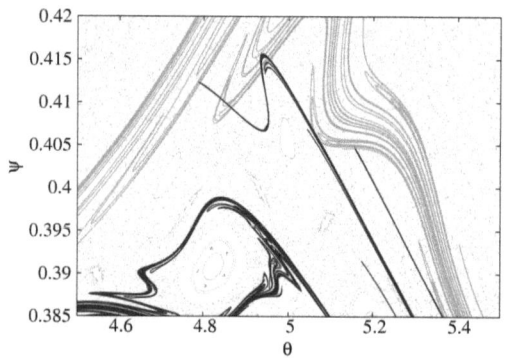 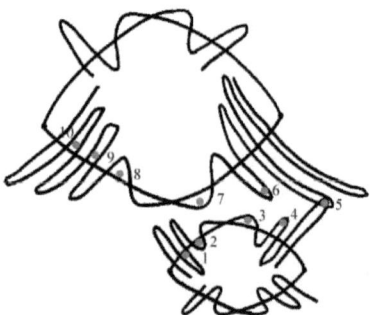

Figure 2.8: Intersection of the unstable manifold of the period 9/5 island chain (black line) and the stable manifold of the period 2/1 island chain (grey line) at a perturbation of $\varepsilon = 4.1/2\pi$.

Figure 2.9: Sketch of the progression of the iteration of a field line within the intersection area of two manifolds.

Increasing the perturbation leads to an increase of the oscillations and the splitting of the manifolds, resulting in an increase of the chaotic area. By further increasing the perturbation, the stable and unstable manifolds of two neighboring island chains start overlapping. Figure 2.8 shows the intersection of the unstable manifold of the period 9/5 island chain (black line) with the stable manifold of the period 2/1 island chain (grey line). Between two neighboring island chains, whose manifolds are not intersecting at the given perturbation level, a transport barrier is located. On the other hand, due to the intersection of the stable manifold of one island chain with the unstable manifold of the other island chain, field lines can iterate from the one to the other and back again, since the area enclosed by the intersecting manifolds belongs to the chaotic areas of both island chains. The chaotic areas around the islands are connected by the intersecting manifolds. So, transport between both island chains occurs as sketched in Fig. 2.9 [13]. In this figure a field line, marked as solid circle, iterates from position 1 to position 10. At position 1 the field line first follows the stable manifold of the small island and then the unstable manifold to position 5, where the unstable manifold of the small island chain intersects with the stable manifold of the large island chain. Then the field line follows the stable, and later the unstable manifold of the large island chain to position 10. Particles, following this field line, are transported from the small island chain to the large island chain. The necessary perturbation for an overlap of the manifolds of neighboring island chains corresponds to the critical one for the break-up of the intact KAM surface in between, determined for the two last intact surfaces in the previous section. Is the perturbation slightly above the critical one, the manifolds of the neighboring island chains intersect only at the very edge of the chaotic layers. This means that field lines in the chaotic layer of the one island chain have to follow its manifolds for a very long time, until they reach the area of intersection and are transported to the other island chain. It will take very long iteration time to see relevant transport effects. Increasing the perturbation also increases the areas of intersection so that the transport from one island chain to the other occurs much faster. This clearly explains the developing of the MSD, shown in Figs. 2.4 and 2.5.

So, the appearance of chaotic motion and transport can be explained by the stable and unstable manifolds. Also the existence of transport barriers and their destruction with increasing perturbation can be explained by the overlapping of the stable manifolds with the unstable ones. The manifolds are playing a fundamental role in understanding and influencing of chaotic transport.

2.8 The question of spontaneously inverted q-profiles

Misguich et al. [29] reported for the non-symmetric tokamap an inversion of the q-profile near the magnetic axis due to the perturbation. However, our calculations show that the q-profile remains equal to one inside the main magnetic island around the magnetic axis, and increases outside. So there is no inversion of the q-profile near the magnetic axis. This results from precise numerical calculations. It is valid for both, the symmetric and the non-symmetric tokamap. Figure 2.10 shows several island chains of the non-symmetric tokamap. Some of them are inside the main island core around the magnetic axis, with $q = 1$. In the figure the periodicities of the island chains are marked. One can see that inside the main island core the toroidal periodicity is equal to the poloidal one, meaning that $q = 1$ is valid inside the whole island. Outside the island, the q-profile increases with the distance to the magnetic axis. So there is no spontaneous inversion of the q-profile.

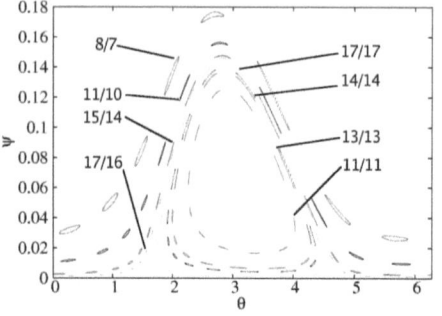
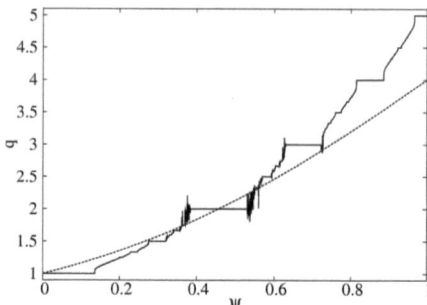

Figure 2.10: Several island chains of the non-symmetric tokamap at a perturbation level of $\varepsilon = 4.5/2\pi$.

Figure 2.11: The q-profile of the non-symmetric tokamap for $\varepsilon = 4.5/2\pi$. Shown are the q-values along the path $\theta = \pi$ starting from $\psi = 0.06$. The broken line depicts the unperturbed q-profile.

Figure 2.11 shows the perturbed q-profile, solid line, of the non-symmetric tokamap for $\varepsilon = 4.5/2\pi$ along the $\theta = \pi$ axis from $\psi = 0.06$ up to $\psi = 1$, compared to the unperturbed q-profile, Eq. (2.3.9), given by the dashed line. The perturbed q-profile is still a monotonic one, while q remains constant inside of islands. The fluctuations at the edges of the constant parts are caused by the chaotic layer, which surrounds the islands. Within these chaotic layers, q, given by the fraction of the toroidal and poloidal periodicities, is no longer numerically calculatable, because the periods are infinite.

2.9 The symmetric revtokamap

To study magnetic shear effects one has to apply a (zeroth-order) non-monotonic q-profile as, e.g., presented in (2.3.10). This leads to the so called revtokamap. In the Figs. 2.12 and 2.13 the symmetric revtokamap and the non-symmetric revtokamap are plotted, respectively.

First, both forms of the revtokamap show differences between each other similar to those already discussed for both forms of the tokamap. But, secondly, both forms of the revtokamap show significant differences to the tokamap with a monotonic q-profile. In the following we compare the symmetric forms of the tokamap and the revtokamap. The main difference between them is the following. When we assume a wall at the position $\psi = 1$, field lines are terminated when they reach $\psi = 1$. In this case the system can be considered as a chaotic scattering system [31]. At approximately $\psi = 0.5$, the main chaotic areas of the tokamap are located around the 2/1 island chain. Close to the wall there are still intact KAM-surfaces, as one can

2.9 The symmetric revtokamap

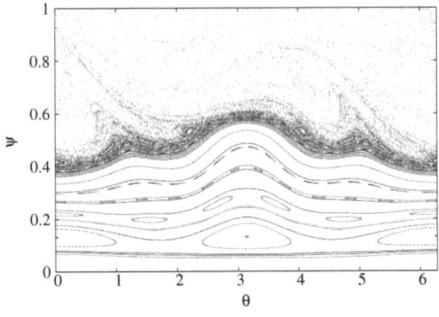
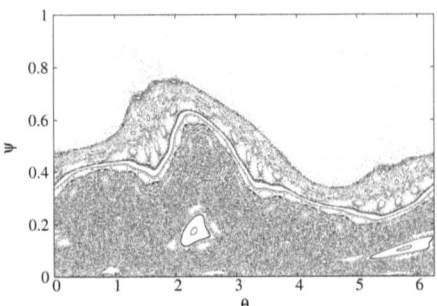

Figure 2.12: Plot of the symmetric revtokamap with a non-monotonic q-profile and $\varepsilon = 6/2\pi$.

Figure 2.13: Same as Fig. 2.12, but for the non-symmetric revtokamap.

see in Fig. 2.1. So the field lines in the chaotic region can never reach the wall for perturbations less than the critical one for the break up of the last upper KAM surface, see Sec. 2.4. The tokamap is typically a closed chaotic system. For the revtokamap the chaotic area is located at the edge of the map, while the center remains very regular, even for very large perturbations $\varepsilon \approx 1$. As seen from Figs. 2.12 and 2.13, there is a magnetic transport barrier for field lines located near the shearless curve $\psi_m \approx 0.55$, where the minimum of the q-profile is reached. The chaotic field lines above can hit the wall. So the revtokamap is an open chaotic system at the edge with a concealed very regular interior. Such a configuration is much more typical for tokamak fusion machines, although a monotonic q-profile, like for the tokamap regime, is used there. Therefore, for the revtokamap it is interesting to analyze the structure of the chaotic area according to the number of toroidal and poloidal rotations of the field lines, until they connect the wall through this area. These so called laminar plots are colored contour plots, where areas with the same rotation numbers are colored in the same way. In chaotic scattering systems a trajectory may leave the system in one of several different ways. The areas of initial coordinates corresponding to the various exit ways are separated by a boundary, which may be fractal [31, 32]. The set of initial conditions for which trajectories leave the system in a particular way is called the basin of the particular mode [32]. Figures 2.14 and 2.15 show these basins for the toroidal and poloidal rotation numbers, respectively.

From the laminar plots we can detect, how many rotations are performed for connecting a point at the wall with the next intersection with the wall. It is interesting to see, how deep field lines can penetrate into the plasma, even when they are escaping quickly to the wall. Figures 2.14 and 2.15 show the fractal structures of the basins in the (θ, ψ)-plane, characterizing wall to wall connections through n toroidal or m poloidal turns, respectively. The symmetric structure of the laminar plots can be understood by distinguishing the transport to the wall during backward and forward iterations separately. The back-iteration of the sym-

Chapter 2. Formation of chaos in the symmetric tokamap regime

metric revtokamap is shown in Fig. 2.16. Here the basins for toroidal turns of magnetic field lines (until they hit the wall) are shown. Together with the corresponding forward-iteration, see Fig. 2.17, it reproduces in the superposition Fig. 2.14 for the total rotation numbers for wall-to-wall connections. It is now interesting to see that, in the case of Fig. 2.16, the border lines are given by the unstable manifolds of the hyperbolic points of the last island chain. Remember, the unstable manifold is defined as the set of points, which converge under the map towards the hyperbolic point, when time goes to $-\infty$. This causes that under the back-iterating map points on the unstable manifold will never hit the wall, resulting in infinite toroidal and poloidal rotation numbers. Similarly, the related stable manifolds characterize the laminar plot for forward-iteration in the case of Fig. 2.17. Note, some parts of the shown manifolds, e.g. upper left corner of Fig. 2.16, have already crossed the wall, causing that there are no further structures. If we shift the wall to higher values of ψ, we would become further basins and further borders.

From here we can also recognize the importance of the stable and unstable manifolds for the dynamics at the edge of open chaotic systems. This will be further investigated on more realistic models in the next sections.

2.9 The symmetric revtokamap

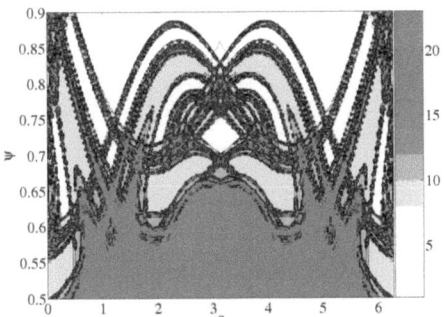

Figure 2.14: Laminar plot of the symmetric revtokamap for the toroidal rotation numbers at $\varepsilon = 6/2\pi$.

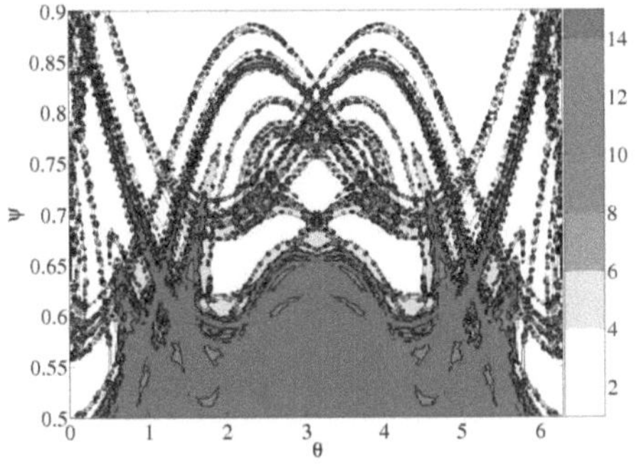

Figure 2.15: Laminar plot of the symmetric revtokamap for the poloidal rotation numbers at $\varepsilon = 6/2\pi$.

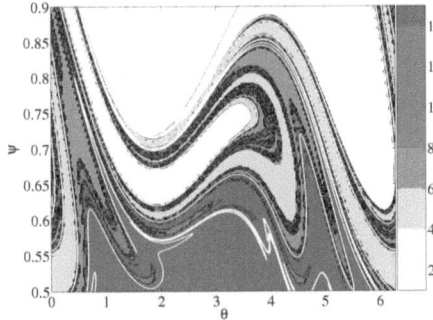
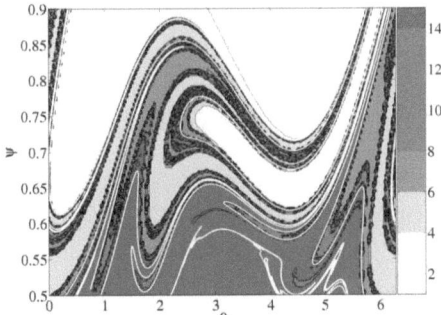

Figure 2.16: Part of the laminar plot, Fig. 2.14, is here analyzed by back-iterating the symmetric revtokamap and counting the toroidal rotations until a line hits the wall at $\psi = 1$. The white line shows the unstable manifold of an hyperbolic point of the last island chain.

Figure 2.17: Same as Fig. 2.16, but for the forward-iteration of the symmetric revtokamap. Now the white line depicts the stable manifold of an hyperbolic point of the last island chain.

Chapter 2. Formation of chaos in the symmetric tokamap regime

Chapter 3

Cylindrical model for magnetic field lines in TEXTOR-DED

An existing tokamak fusion experiment is the TEXTOR, Torus EXperiment for Technology Oriented Research, of the Forschungszentrum Jülich. The dynamic ergodic divertor [2, 3], DED, is a system of 16 helical coils, mounted at the inner side of the TEXTOR. The torus has the major radius $R_0 = 1.75$ m and the minor radius $r_w = 0.477$ m, the radial position of the wall, while the coils are installed at a minor radius of $r_c = 0.5325$ m. Each coil passes during one single toroidal rotation $0 \leq \varphi \leq 2\pi$, with the toroidal angle φ, the poloidal angle area $\pi - \theta_c \leq \theta \leq \pi + \theta_c$ with the poloidal angle θ. Thereby, $\theta = \pi$ represents the inner side of the torus and $\theta_c = \pi/5$ is the half width of the coil area. The configuration is sketched in Fig. 3.1. Each coil can be connected to current separately, whereas 15 kA is the maximum current for each coil. Due to the current, a magnetic field is created [5], dealing as a perturbation field for the main magnetic field, which consists of a strong constant field B_0 in toroidal direction and a poloidal field due to the plasma current, shown in Sec. 3.3. Because of the DED field, an area is created at the torus edge, where the field lines are showing chaotic behavior, the so called ergodic region. The purpose of this controlled perturbation is to influence and manage the heat and particle transport to the wall. The DED system is an open chaotic system with a regular interior, similar to the revtokamap, but, as we shall see later, with a monotonic q-profile. Therefore, we can directly apply the results of the previous chapter to the DED system.

We start our analysis of the DED and its field in a more simplified cylindrical model, shown in Fig. 3.2. Due to the cylindrical model, we can concentrate on the pure effects of the DED itself. In contrast to the system in toroidal geometry, which was presented by S. Abdullaev et al. in [4, 5], we do not have to consider toroidal effects like the Shafranov shift [33]. The spectrum of the perturbation and the safety factor also remain unchanged in cylindrical geometry. In toroidal geometry, corrections have to be applied [5]. The cylindrical model, which can be calculated completely analytical, is therefore a good approach to study

Chapter 3. Cylindrical model for magnetic field lines in TEXTOR-DED

the effects of the stable and unstable manifolds on the heat and particle transport and their contribution to the formation and developing of the chaotic edge region, caused by the external DED perturbation.

We consider a cylinder with the radius r_c, representing the minor radius of the coils at the torus, and the hight $2\pi R_0$, whereas R_0 represents the major radius of the torus. We are using cylindrical coordinates (r, θ, z) and continue the cylinder periodically in both directions of z to neglect unwanted boundary effects at the top and bottom of the cylinder. The DED coils are attached to the wall of the cylinder in the angle area $\pi - \theta_c \leq \theta \leq \pi + \theta_c$, also continued periodically in both directions of z. We introduce the angle φ with $z = \varphi \cdot R_0$, corresponding to the toroidal angle of the torus.

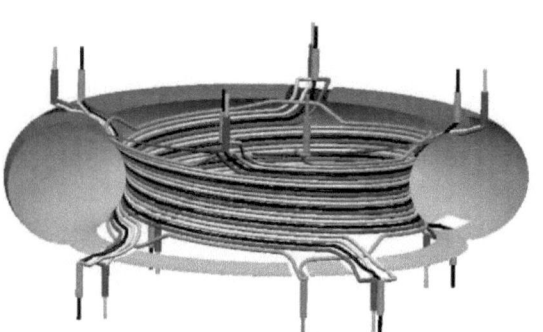

Figure 3.1: Sketch of the DED at TEXTOR

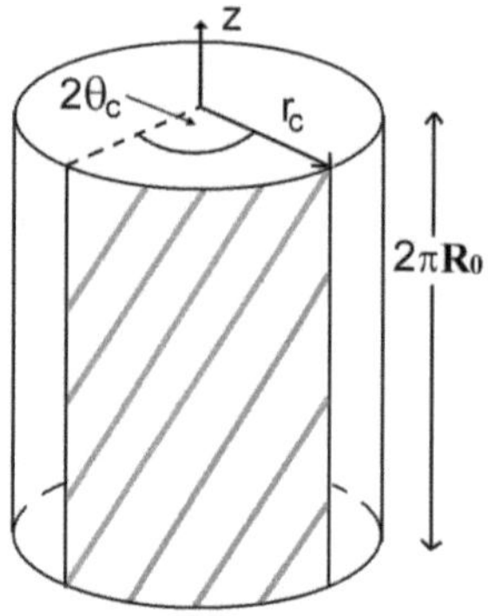

Figure 3.2: Sketch of the cylindrical model used here

The mapping technique (see Sec. 2.2) has proven to be very useful, according to the analysis of the tokamap. Therefore, we will construct a mapping procedure similar to Eqs. (2.2.11)-(2.2.13) for the cylindrical DED model in order to continue the investigation of statistical properties and transport phenomena of chaotic fusion plasma systems. Similar to the tokamap regime we only consider magnetic field lines, assuming that the thermal plasma particles are mainly following the field lines. Note, there is no influence of the particle charge on the external fields, so any particles are considered to be "test-particles" only. We also neglect all further plasma effects, except of the plasma current, which specifies the safety factor. The cylindrical DED model provides us with a symplectic symmetric map for the magnetic field lines of the main helical field, given by the superposition of the constant field B_0 in z-direction and the poloidal field of the plasma current, perturbed by the field of the DED coils.

3.1 The current density

As a first step, we have to determine the magnetic field, created by the DED coils. This field can be derived from the current density in the cylinder surface, which is based on the current distribution. We connect the coils to the following current distribution $I = 0, I_0, 0, -I_0, 0, \ldots$, which means that the j-th coil is carrying the current

$$I_j = I_0 \sin(\frac{j-1}{2}\pi + \omega t), \qquad (3.1.1)$$

which can also rotate in time with the frequency ω, describing the DED's dynamical state of operation. For $\omega = 0$ the DED operates statically. Considering the geometry mentioned above, we can express θ as a function of φ along the coils. We get for the j-th coil

$$\theta_j = \theta_1(\varphi) + \frac{j-1}{8}\theta_c \quad \text{with} \quad \theta_1(\varphi) = \frac{\theta_c}{\pi}\varphi + \theta_1(0). \qquad (3.1.2)$$

The starting point for the first coil is arbitrarily chosen to be $\theta_1(0) = \pi - \theta_c$. The current density

$$J = \sum_{j=1}^{16} \frac{I_j}{r_c}\delta(\theta - \theta_j) = \frac{I_0}{r_c}\sum_j \sin(\frac{j-1}{2}\pi + \omega t)\delta(\theta - \frac{\theta_c}{\pi}\varphi - \theta_1(0) - \frac{j-1}{8}\theta_c) \qquad (3.1.3)$$

follows as the sum over all single currents, taken at their specific angle position, which is set by the δ-function here. We extend the summation to infinity

$$J = \frac{I_0}{r_c}g(\theta)\sum_{j=-\infty}^{\infty}\sin(\frac{j}{2}\pi + \omega t)\delta(\theta - \frac{\theta_c}{\pi}\varphi - \theta_1(0) - \frac{j}{8}\theta_c), \qquad (3.1.4)$$

using the step-function

$$g(\theta) = \begin{cases} 1, & \pi - \theta_c \leq \theta \leq \pi + \theta_c \\ 0, & \text{else} \end{cases}, \qquad (3.1.5)$$

which specifies the correct angle area of the coils. This correctly describes the necessary periodic continuation of the cylinder. We define

$$n = (\theta - \frac{\theta_c}{\pi}\varphi - \theta_1(0))\frac{8}{\theta_c} \qquad (3.1.6)$$

and obtain for the current density

$$J = \frac{8I_0}{\theta_c r_c}g(\theta)\sum_{j=-\infty}^{\infty}\sin(\frac{j}{2}\pi + \omega t)\delta(j - n), \qquad (3.1.7)$$

Chapter 3. Cylindrical model for magnetic field lines in TEXTOR-DED

using the probabilities of the δ-function: $\delta(ax) = \frac{1}{a}\delta(x)$ with $a > 0$ and $\delta(-x) = \delta(x)$. The form (3.1.7) of the current density is not appropriate, because of the δ-functions. To find a better representation of the current density, we expand the δ-function into its harmonics, using the Poisson summation rule

$$\sum_{j=-\infty}^{\infty} \delta(j-n) = 1 + 2\sum_{k=1}^{\infty} \cos(2\pi k n) \,. \tag{3.1.8}$$

This is presented in Appendix B, because it is a long but straight forward conversion. We find

$$J = \frac{8I_0}{\theta_c r_c} g(\theta) \sum_{k=0}^{\infty} \sin((-1)^k \frac{1}{2}\pi(2k+1)n + \omega t) \,. \tag{3.1.9}$$

We introduce the toroidal and poloidal main mode numbers n_0 and m_0, respectively, by

$$\frac{\theta_c}{\pi} = \frac{n_0}{m_0}, \tag{3.1.10}$$

whereas $n_0 = 4$ and $m_0 = 20$ are valid for the DED. We also define the shortcuts $J_0 := \frac{8I_0}{r_c \theta_c} = I_0 \frac{2m_0}{\pi r_c}$, $(2k+1)m_0 = m_k$ and $(2k+1)n_0 = n_k$. Using n from Eq. (3.1.6), we get

$$J = J_0 g(\theta) \sum_{k=0}^{\infty} (-1)^k \sin\left(m_k \theta - n_k \varphi - m_k \theta_1(0) + (-1)^k \omega t\right) \,. \tag{3.1.11}$$

With $\theta_1(0) = \pi - \theta_c$ and $m_k(\pi - \theta_c) = (m_k - n_k)\pi$ it follows from Eq. (3.1.11)

$$J = J_0 g(\theta) \sum_{k=0}^{\infty} (-1)^{k+m_k-n_k} \sin\left(m_k \theta - n_k \varphi + (-1)^k \omega t\right) , \tag{3.1.12}$$

using the sinus addition theorem. This equation for the current density still includes the step function $g(\theta)$, which specifies the angle area of the coils. By Fourier transforming Eq. (3.1.12) with respect to θ, we can expand

$$g(\theta) \sin\left(m_k \theta - n_k \varphi + (-1)^k \omega t\right)$$

into its harmonics, to finally find the proper description of the current density. We introduce the shortcut $\chi := -n_k \varphi + (-1)^k \omega t$. Then the current density (3.1.12) reads in Fourier representation

$$J = \sum_{m=-\infty}^{\infty} \sum_{k=0}^{\infty} (-1)^{k+m_k-n_k} J_0 \frac{1}{2\pi} \int_{-\infty}^{\infty} g(\theta) \sin(m_k \theta + \chi) e^{im\theta} \, d\theta \, e^{-im\theta} \,. \tag{3.1.13}$$

3.1 The current density

Now we consider the sum over m and the integral. The step function $g(\theta)$ constricts the integration limits so that we obtain

$$\sum_m \int_{\pi-\theta_c}^{\pi+\theta_c} \sin(m_k\theta + \chi) e^{im\theta} \, d\theta \, e^{-im\theta}$$

$$= \sum_m \int_{\pi-\theta_c}^{\pi+\theta_c} \frac{1}{2i} \left(e^{i(m_k\theta+\chi)} - e^{-i(m_k\theta+\chi)} \right) e^{im\theta} \, d\theta \, e^{-im\theta}$$

$$= \sum_m \frac{1}{2i} e^{-i(m\theta-\chi)} \int_{\pi-\theta_c}^{\pi+\theta_c} e^{i(m+m_k)\theta} \, d\theta - \sum_m \frac{1}{2i} e^{-i(m\theta+\chi)} \int_{\pi-\theta_c}^{\pi+\theta_c} e^{i(m-m_k)\theta} \, d\theta \ .$$

In the first sum we change the summation index $m \to -m$. Because m goes from $-\infty$ to ∞, the sum is not affected. We obtain furthermore

$$\sum_m \frac{1}{2i} e^{i(m\theta+\chi)} \int_{\pi-\theta_c}^{\pi+\theta_c} e^{-i(m-m_k)\theta} \, d\theta - \sum_m \frac{1}{2i} e^{-i(m\theta+\chi)} \int_{\pi-\theta_c}^{\pi+\theta_c} e^{i(m-m_k)\theta} \, d\theta$$

$$= \sum_m \frac{1}{2i} e^{i(m\theta+\chi)} \frac{1}{i(m-m_k)} \left(-e^{-i(m-m_k)(\pi+\theta_c)} + e^{-i(m-m_k)(\pi-\theta_c)} \right)$$

$$- \sum_m \frac{1}{2i} e^{-i(m\theta+\chi)} \frac{1}{i(m-m_k)} \left(e^{i(m-m_k)(\pi+\theta_c)} - e^{i(m-m_k)(\pi-\theta_c)} \right)$$

$$= \sum_m \frac{1}{2i} \frac{1}{i(m-m_k)} \left[e^{i(m\theta+\chi)} e^{-i(m-m_k)\pi} \left(-e^{-i(m-m_k)\theta_c} + e^{i(m-m_k)\theta_c} \right) \right.$$
$$\left. - e^{-i(m\theta+\chi)} e^{i(m-m_k)\pi} \left(e^{i(m-m_k)\theta_c} - e^{-i(m-m_k)\theta_c} \right) \right]$$

$$= \sum_m \frac{2\sin((m-m_k)\theta_c)}{m-m_k} \frac{1}{2i} \left[e^{i(m\theta+\chi)} e^{-i(m-m_k)\pi} - e^{-i(m\theta+\chi)} e^{i(m-m_k)\pi} \right]$$

$$= \sum_m (-1)^{m-m_k} \frac{2\sin((m-m_k)\theta_c)}{m-m_k} \sin(m\theta + \chi) \ .$$

Inserting this result into Eq. (3.1.13), we get the final formula for the current density

$$J = \sum_{m=-\infty}^{\infty} \sum_{k=0}^{\infty} J_m^{(k)} \sin(m\theta - n_k\varphi + (-1)^k \omega t) \ , \qquad (3.1.14)$$

with $J_m^{(k)} = (-1)^{k+m_k-n_k} J_0 g_m^{(k)}$. The Fourier spectrum $g_m^{(k)}$ is given by

$$g_m^{(k)} = (-1)^{m-m_k} \frac{\sin((m-m_k)\theta_c)}{(m-m_k)\pi} \ . \qquad (3.1.15)$$

Chapter 3. Cylindrical model for magnetic field lines in TEXTOR-DED

3.2 The magnetic field of the DED coils

From the current density (3.1.14) we can calculate the magnetic field of the DED coils. For this we use the magnetic scalar potential $\phi(r, \theta, z)$, related to the magnetic field by

$$\vec{B} = -\nabla \phi . \qquad (3.2.1)$$

Inside the cylinder no currents are present. We can define the scalar potential there since outside the coils, which are carrying the current, $\nabla \times \vec{B} = 0$ is valid. Using the Maxwell equation $\nabla \cdot \vec{B} = 0$, we can derive the Laplace equation

$$\Delta \phi = 0 \quad \text{with} \quad \Delta = \frac{1}{r}\frac{\partial}{\partial r}(r\frac{\partial}{\partial r}) + \frac{1}{r^2}\frac{\partial^2}{\partial \theta^2} + \frac{\partial^2}{\partial z^2} \qquad (3.2.2)$$

from Eq. (3.2.1). We solve the Laplace equation by factorization [34]

$$\phi(r, \theta, z) = f(r)g(\theta)h(z) \qquad (3.2.3)$$

and get

$$gh\frac{1}{r}\frac{\partial}{\partial r}(r\frac{\partial}{\partial r})f + fh\frac{1}{r^2}\frac{\partial^2}{\partial \theta^2}g + fg\frac{\partial^2}{\partial z^2}h = 0 \qquad (3.2.4)$$

$$\Leftrightarrow \frac{1}{f}\frac{1}{r}\frac{\partial}{\partial r}(r\frac{\partial}{\partial r})f + \frac{1}{g}\frac{1}{r^2}\frac{\partial^2}{\partial \theta^2}g = -\frac{1}{h}\frac{\partial^2}{\partial z^2}h . \qquad (3.2.5)$$

Each side of this equation depends on different variables. To fulfill the equation generally, both sides have to be constant. Then we find for $h(z)$

$$-\frac{1}{h}\frac{\partial^2}{\partial z^2}h = c_1^2 \quad \Leftrightarrow \quad \frac{\partial^2}{\partial z^2}h = -c_1^2 h \qquad (3.2.6)$$

$$\Rightarrow h(z) = c\sin(c_1 z) + d\cos(c_1 z) . \qquad (3.2.7)$$

The left hand-side of Eq. (3.2.5) results in

$$\frac{r}{f}\frac{\partial}{\partial r}(r\frac{\partial}{\partial r})f - c_1^2 r^2 = -\frac{1}{g}\frac{\partial^2}{\partial \theta^2}g , \qquad (3.2.8)$$

whereas again both sides depend on different variables, so they are also constant. Therefore, we find for $g(\theta)$

$$-\frac{1}{g}\frac{\partial^2}{\partial \theta^2}g = c_2^2 \quad \Leftrightarrow \quad \frac{\partial^2}{\partial \theta^2}g = -c_2^2 g \qquad (3.2.9)$$

$$\Rightarrow g(\theta) = a\sin(c_2 \theta) + b\cos(c_2 \theta) \qquad (3.2.10)$$

3.2 The magnetic field of the DED coils

and the differential equation

$$\left[r^2 \frac{\partial^2}{\partial r^2} + r \frac{\partial}{\partial r} - c_1^2 r^2 - c_2^2 \right] f(r) = 0 \tag{3.2.11}$$

for $f(r)$, which is solved by the modified Bessel functions I and K. We can construct the solution

$$f(r) = \begin{cases} A I_{c_2}(c_1 r) & , r < r_c \\ B K_{c_2}(c_1 r) & , r > r_c \end{cases} \tag{3.2.12}$$

for Eq. (3.2.11). This is the form of the solution for f, because K diverges for $r \to 0$ and therefore only I is a proper solution inside the cylinder. Outside of the cylinder the field must decrease with increasing r, so only K can be a solution there. Both solutions are divided by the current carrying surface of the cylinder at $r = r_c$, where the coils are located. We obtain the ansatz for ϕ

$$\phi(r, \theta, z) = \begin{cases} A I_{c_2}(c_1 r)(a \sin(c_2\theta) + b \cos(c_2\theta))(c \sin(c_1 z) + d \cos(c_1 z)) & , r < r_c \\ B K_{c_2}(c_1 r)(a \sin(c_2\theta) + b \cos(c_2\theta))(c \sin(c_1 z) + d \cos(c_1 z)) & , r > r_c \end{cases} . \tag{3.2.13}$$

To determine the exact solution, we have to take the boundary conditions into account. Due to the periodicity in θ and z, we demand

$$\theta + 2\pi = \theta \quad \Rightarrow \quad c_2 \in \mathbb{Z} \tag{3.2.14}$$

$$z + 2\pi R_0 = z \quad \Rightarrow \quad c_1 \cdot R_0 \in \mathbb{Z} \tag{3.2.15}$$

to be valid for the constants. Also we demand

$$\vec{e}_r \times (\vec{B}_{r=r_c+0} - \vec{B}_{r=r_c-0}) = \mu_0 \vec{J} \tag{3.2.16}$$

at $r = r_c$ with the current density vector $\vec{J} = (0, J \sin \alpha, J \cos \alpha)$, whereas α is given by the geometrical relation

$$\tan \alpha = \frac{n_0 r_c}{m_0 R_0}, \tag{3.2.17}$$

and J is the current density (3.1.14). From Eq. (3.2.16) we get two equations

$$-B_{z,\, r=r_c+0} + B_{z,\, r=r_c-0} = \mu_0 J \sin \alpha, \tag{3.2.18}$$

$$B_{\theta,\, r=r_c+0} - B_{\theta,\, r=r_c-0} = \mu_0 J \cos \alpha, \tag{3.2.19}$$

which involve the current density. The third equation

$$B_{r,\, r=r_c+0} = B_{r,\, r=r_c-0} \tag{3.2.20}$$

Chapter 3. Cylindrical model for magnetic field lines in TEXTOR-DED

represents the continuity of the radial component of the magnetic field at $r = r_c$. Using Eq. (3.2.13) and $\vec{B} = -\nabla \phi$, it follows from Eqs. (3.2.18) and (3.2.19)

$$c_1(a\sin(c_2\theta) + b\cos(c_2\theta))(c\cos(c_1z) - d\sin(c_1z))(BK_{c_2}(c_1r_c) - AI_{c_2}(c_1r_c))$$
$$= \mu_0 \sin\alpha J_m^{(k)} \sin(m\theta - (2k+1)n_0\varphi + (-1)^k\omega t), \qquad (3.2.21)$$

$$-\frac{c_2}{r_c}(a\cos(c_2\theta) - b\sin(c_2\theta))(c\sin(c_1z) + d\cos(c_1z))(BK_{c_2}(c_1r_c) - AI_{c_2}(c_1r_c))$$
$$= \mu_0 \cos\alpha J_m^{(k)} \sin(m\theta - (2k+1)n_0\varphi + (-1)^k\omega t). \qquad (3.2.22)$$

for each single summand of the current density J, given by Eq. (3.1.14). We can identify the constants

$$c_2 = m \quad \text{and} \quad c_1 = \frac{(2k+1)n_0}{R_0} = \frac{n_k}{R_0}, \qquad (3.2.23)$$

which fulfill Eqs. (3.2.14) and (3.2.15).

Because Eqs. (3.2.21) and (3.2.22) have to be valid for all values of θ and φ, the particular terms on both sides of the equations must cancel each other. Therefore, it follows from Eq. (3.2.21)

$$ac\sin(m\theta)\cos(n_k\varphi) + bc\cos(m\theta)\cos(n_k\varphi)$$
$$-ad\sin(m\theta)\sin(n_k\varphi) - bd\cos(m\theta)\sin(n_k\varphi)$$
$$= \sin(m\theta - n_k\varphi + (-1)^k\omega t)$$
$$= \cos((-1)^k\omega t)\sin(m\theta)\cos(n_k\varphi) - \cos((-1)^k\omega t)\cos(m\theta)\sin(n_k\varphi)$$
$$+ \sin((-1)^k\omega t)\cos(m\theta)\cos(n_k\varphi) + \sin((-1)^k\omega t)\sin(m\theta)\sin(n_k\varphi)$$
$$\Rightarrow \quad ac = bd = \cos((-1)^k\omega t) \quad bc = -ad = \sin((-1)^k\omega t) \qquad (3.2.24)$$

and from Eq. (3.2.22)

$$-ac\cos(m\theta)\sin(n_k\varphi) - bc\sin(m\theta)\sin(n_k\varphi)$$
$$+ad\cos(m\theta)\cos(n_k\varphi) - bd\sin(m\theta)\cos(n_k\varphi)$$
$$= \cos((-1)^k\omega t)\sin(m\theta)\cos(n_k\varphi) - \cos((-1)^k\omega t)\cos(m\theta)\sin(n_k\varphi)$$
$$+ \sin((-1)^k\omega t)\cos(m\theta)\cos(n_k\varphi) + \sin((-1)^k\omega t)\sin(m\theta)\sin(n_k\varphi)$$
$$\Rightarrow \quad ac = bd = \cos((-1)^k\omega t) \quad bc = -ad = \sin((-1)^k\omega t). \qquad (3.2.25)$$

Using these results we get in Eq. (3.2.13)

$$(a\sin(c_2\theta) + b\cos(c_2\theta))(c\sin(c_1z) + d\cos(c_1z)) = \cos(m\theta - n_k\varphi + (-1)^k\omega t) \qquad (3.2.26)$$

3.2 The magnetic field of the DED coils

and therefore for the summands $\phi_m^{(k)}$ of the potential

$$\phi_m^{(k)}(r,\theta,z) = \begin{cases} AI_m(c_1 r)\cos(m\theta - n_k\varphi + (-1)^k\omega t) &, r < r_c \\ BK_m(c_1 r)\cos(m\theta - n_k\varphi + (-1)^k\omega t) &, r > r_c \end{cases}, \quad (3.2.27)$$

while the total potential is then given as the sum over m and k. With this result, Eqs. (3.2.21) and (3.2.22) are simplified to

$$\frac{n_k}{R_0}(BK_m(c_1 r_c) - AI_m(c_1 r_c)) = \mu_0 \sin\alpha J_m^{(k)}, \quad (3.2.28)$$

$$\frac{m}{r_c}(BK_m(c_1 r_c) - AI_m(c_1 r_c)) = \mu_0 \cos\alpha J_m^{(k)}. \quad (3.2.29)$$

It can be shown that Eqs. (3.2.28) and (3.2.29) are equivalent, when they are summed over m. For this we add an additional summand $\delta J_m \neq 0$ to Eq. (3.2.21) and show that

$$\sum_{m=-\infty}^{\infty} \delta J_m = 0 \quad (3.2.30)$$

is valid. Equation (3.2.21) reads then

$$\frac{n_k}{R_0}(BK_m(c_1 r_c) - AI_m(c_1 r_c)) = \mu_0 \sin\alpha (J_m^{(k)} + \delta J_m). \quad (3.2.31)$$

Combining this equation with Eq. (3.2.22), we obtain

$$\frac{n_k r_c}{R_0 m} = (1 + \frac{\delta J_m}{J_m^{(k)}})\tan\alpha. \quad (3.2.32)$$

Using Eq. (3.2.17) and $J_m^{(k)} = (-1)^{k+m-n_k} J_0 \frac{\sin((m-m_k)\theta_c)}{(m-m_k)\pi}$, we find

$$\begin{aligned}
\delta J_m &= (\frac{m_k}{m} - 1)J_m^{(k)} & (3.2.33) \\
&= J_0 \frac{1}{\pi}(-1)^{k+m-n_k+1}\frac{\sin((m-m_k)\theta_c)}{m} & (3.2.34) \\
&= J_0 \frac{n_0}{m_0}(-1)^{k+m-n_k+1}\left[\frac{\sin(m\theta_c)}{m\theta_c}\cos(n_k\pi) - \frac{\cos(m\theta_c)}{m\theta_c}\sin(n_k\pi)\right] & (3.2.35) \\
&= J_0 \frac{n_0}{m_0}(-1)^{k+m+1}\frac{\sin(m\theta_c)}{m\theta_c}. & (3.2.36)
\end{aligned}$$

It can easily be seen that

$$\sum_{m=-\infty}^{\infty} \delta J_m = J_0 \frac{n_0}{m_0}(-1)^{k+1}\sum_{m=-\infty}^{\infty}(-1)^m \frac{\sin(m\theta_c)}{m\theta_c} = 0, \quad (3.2.37)$$

Chapter 3. Cylindrical model for magnetic field lines in TEXTOR-DED

which proves that Eqs. (3.2.31) and (3.2.22) are equivalent, when summed over m.

In the following we use Eqs. (3.2.29) and (3.2.20) to determine the constants A and B. Combining both equations, we get

$$I'_m(c_1 r_c)\left(-\frac{r_c}{m}\mu_0 \cos\alpha J_m^{(k)} + BK_m(c_1 r_c)\right) = BI_m(c_1 r_c)K'_m(c_1 r_c)$$
$$\Leftrightarrow B[I_m(c_1 r_c)K'_m(c_1 r_c) - I'_m(c_1 r_c)K_m(c_1 r_c)] = -I'_m(c_1 r_c)\frac{r_c}{m}\mu_0 \cos\alpha J_m^{(k)}$$

for B, and similar for A

$$A[I_m(c_1 r_c)K'_m(c_1 r_c) - I'_m(c_1 r_c)K_m(c_1 r_c)] = K'_m(c_1 r_c)\frac{r_c}{m}\mu_0 \cos\alpha J_m^{(k)} \ .$$

We still have to calculate $I_m(x)K'_m(x) - I'_m(x)K_m(x)$ with $x = c_1 r_c = \frac{n_k r_c}{R_0}$, but, as can be seen in [35], this is the Wronskian determinant of modified Bessel functions, which reads

$$I_m(x)K'_m(x) - I'_m(x)K_m(x) = \frac{1}{x} \ . \tag{3.2.38}$$

Finally, we get for the potential inside the cylinder

$$\phi(r,\theta,\varphi) = \sum_{k=0}^{\infty}\sum_{m=-\infty}^{\infty}\phi_{m,k}I_m(\frac{n_k r}{R_0})\cos(m\theta - n_k\varphi + (-1)^k\omega t) \tag{3.2.39}$$

with

$$\phi_{m,k} = A = K'_m(\frac{n_k r_c}{R_0})\frac{n_k r_c^2}{R_0 m}\mu_0 \cos\alpha J_m^{(k)} \tag{3.2.40}$$
$$= (-1)^{k+m-n_k}J_0\mu_0 K'_m(\frac{n_k r_c}{R_0})\frac{n_k r_c^2}{R_0 m}\frac{\sin((m-m_k)\theta_c)}{(m-m_k)\pi}\cos\alpha \ . \tag{3.2.41}$$

In the following it is sufficient to take only the $k = 0$ mode into account, because all higher modes can be neglected compared to $k = 0$. The $m = m_k$ mode contributes most for a fixed k, as one can see from Eq. (3.1.15). So, the contribution of each k-mode is mainly given by the factor $K'_{m_k}(\frac{n_k r_c}{R_0})I_{m_k}(\frac{n_k r}{R_0})$. Under the conditions $x \ll n$ and $n > 0$, modified Bessel functions can be approximated by

$$I_n(x) \approx \frac{1}{n!}\left(\frac{x}{2}\right)^n \quad K_n(x) \approx \frac{(n-1)!}{2}\left(\frac{x}{2}\right)^{-n} \ . \tag{3.2.42}$$

Because here $\frac{n_k r_c}{R_0} \ll m_k$ is valid for all values of k, we can approximate

$$K'_{m_k}(\frac{n_k r_c}{R_0})I_{m_k}(\frac{n_k r}{R_0}) \approx -\frac{R_0}{2r_c n_k}\left(\frac{r}{r_c}\right)^{(2k+1)m_0} \ . \tag{3.2.43}$$

For $k \geq 1$ this factor is very small compared to $k = 0$. So, only the $k = 0$ mode has to be taken into account. Then the scalar potential reads

$$\phi(r,\theta,\varphi) = \sum_{m=-\infty}^{\infty} \phi_m I_m(\frac{n_0 r}{R_0}) \cos(m\theta - n_0\varphi + wt) ,\qquad(3.2.44)$$

and the magnetic field is given by $\vec{B} = -\nabla\phi$.

3.3 The safety factor

The magnetic field created by the helical DED coil system deals as a perturbation for the main magnetic field, which is given by the superposition of a constant field B_0 in z direction and the magnetic field of the plasma current $I_p(r)$. This superposition creates the helical main magnetic field

$$\vec{B}_g = B_0 \vec{e}_z + \frac{\mu_0 I_p(r)}{2\pi r} \vec{e}_\theta .\qquad(3.3.1)$$

The plasma current $I_p(r)$ is the important parameter for the main field and is directly related to the safety factor $q(r)$. $I_p(r)$ is the amount of current running through a disk with radius r perpendicular to the current direction.

Because the mapping technique is based on a Hamiltonian description, we need to derive the latter from the magnetic field. At first we concentrate on the Hamiltonian for the main field only. For this we use the Clebsch form [25]

$$\vec{B}_g = \nabla\psi \times \nabla\theta - \nabla H_0 \times \nabla\varphi = \begin{pmatrix} \frac{\partial\psi}{\partial r} \\ \frac{1}{r}\frac{\partial\psi}{\partial\theta} \\ \frac{\partial\psi}{\partial z} \end{pmatrix} \times \begin{pmatrix} 0 \\ \frac{1}{r} \\ 0 \end{pmatrix} - \begin{pmatrix} \frac{\partial H_0}{\partial r} \\ \frac{1}{r}\frac{\partial H_0}{\partial\theta} \\ \frac{\partial H_0}{\partial z} \end{pmatrix} \times \begin{pmatrix} 0 \\ 0 \\ \frac{1}{R_0} \end{pmatrix} \qquad(3.3.2)$$

of the magnetic field. The toroidal flux ψ and the poloidal angle θ are used as canonical variables. The poloidal flux H_0 is used as Hamiltonian, while the toroidal angle φ deals as time-like variable. The Clebsch form was already introduced in Sec. 2.1 for the tokamap. Knowing the magnetic field B_g, we determine the fluxes ψ and H_0 so that Eq. (3.3.2) is fulfilled. Using the equation for the radial component, we obtain

$$0 = -\frac{1}{R_0}\frac{\partial H_0}{\partial\theta} \Rightarrow H_0(r,\theta,\varphi) = H_0(r,\varphi) .\qquad(3.3.3)$$

Then we can determine the Hamiltonian from the poloidal component

$$\frac{\mu_0 I_p(r)}{2\pi r} = \frac{1}{R_0}\frac{\partial H_0}{\partial r} \Rightarrow H_0 = \frac{\mu_0 R_0}{2\pi}\int \frac{I_p(r)}{r} dr .\qquad(3.3.4)$$

Chapter 3. Cylindrical model for magnetic field lines in TEXTOR-DED

The z-component specifies the toroidal flux

$$rB_0 = \frac{\partial \psi}{\partial r} \quad \Rightarrow \quad \psi = \frac{1}{2}B_0 r^2 \quad \Leftrightarrow \quad r = \sqrt{\frac{2\psi}{B_0}}. \tag{3.3.5}$$

Now we can derive the relation between the plasma current and the safety factor, given by the main Hamiltonian

$$H_0(\psi) = \int \frac{d\psi}{q(\psi)}. \tag{3.3.6}$$

We obtain

$$q(r) = \frac{2\pi B_0}{\mu_0 R_0} \frac{r^2}{I_p(r)}. \tag{3.3.7}$$

The shape of the q-profile depends on the shape of the plasma current. We use

$$I_p(r) = I_p \left[1 - \left(1 - \frac{r^2}{a^2}\right)^\nu\right] \tag{3.3.8}$$

for the shape of the plasma current, with the total current $I_p = I_p(a)$ at the plasma edge $r = a$. Outside of the plasma, $r > a$, the plasma current $I_p(r)$ is constant and equal to I_p. The exponent ν is given by the ratio of the q-profile at the edge $q(a) = q_a$ to the q-profile at the center $q(0) = q_0$

$$\nu = \frac{q_a}{q_0}. \tag{3.3.9}$$

With

$$q_a = \frac{2\pi B_0 a^2}{\mu_0 R_0 I_p} \tag{3.3.10}$$

we obtain

$$q(r) = q_a \frac{r^2/a^2}{1 - \left(1 - \frac{r^2}{a^2}\right)^\nu} \quad \text{for} \quad r \leq a,$$
$$q(r) = q_a \frac{r^2}{a^2} \quad \text{for} \quad r > a. \tag{3.3.11}$$

Figure 3.3 shows the shape of the q-profile, which we are using for the DED model, for two different parameter configurations. The first one, given by the solid line, is valid for the situation, when the plasma fills the entire inside of the cylinder, meaning that the radius of the plasma is equal to the radius of the cylinder. For this profile the resonant surface $q = 3$ is located at $r = 0.43$ m. This profile will be used for the cylindrical model and is called q-profile "a" in the following. The second profile, profile "b", given by the dashed line, is more typical for the real toroidal experiment, where the plasma radius a is smaller than the minor radius of the torus. The center of the plasma can also be shifted compared to the geometrical center. Profile "b" will be used later. An important fact is that the q-profile is always monotonic. In

3.3 The safety factor

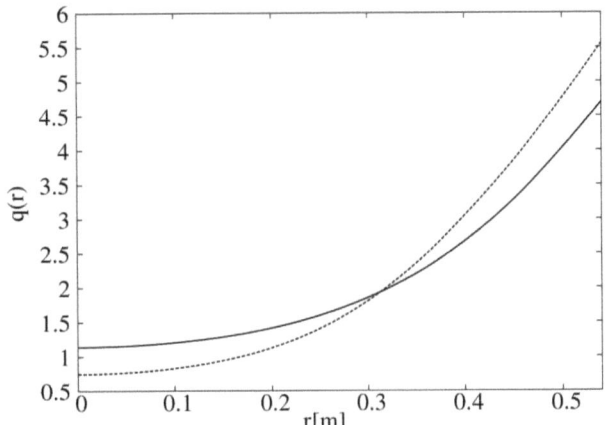

Figure 3.3: q-profiles for two different sets of parameters. Solid line: $a = 0.5325$ m, $I_p = 390$ kA $\Rightarrow q_a = 4.57$, $\nu = 4 \Rightarrow q_0 = 1.14$. Dashed line: $a = 0.46$ m, $I_p = 330$ kA $\Rightarrow q_a = 4$, $\nu = 5.37 \Rightarrow q_0 = 0.75$. For both is $B_0 = 2.2$ T and $R_0 = 1.75$ m.

contrast to the tokamap, where we used a non-monotonic reversed shear profile to create an open chaotic system, here we get an open chaotic system with a monotonic q-profile, as we will see in the next sections.

The derivative of the main Hamiltonian H_0 with respect to the action variable ψ is used in Eq. (2.2.12) of the iteration procedure. According to Eq. (3.3.6), we define the winding number

$$\Omega(\psi) := \frac{\partial H_0}{\partial \psi} = \frac{1}{q(\psi)} . \tag{3.3.12}$$

For the q-profile "a", the solid line of Fig. 3.3 with $\nu = 4$, we obtain for Ω and its first derivative with respect to ψ

$$\Omega(\psi) = \frac{1}{q_a}\left(4 - 6\frac{\psi}{\psi_a} + 4\frac{\psi^2}{\psi_a^2} - \frac{\psi^3}{\psi_a^3}\right) \tag{3.3.13}$$

$$\Omega'(\psi) = \frac{1}{q_a \psi_a}\left(-6 + 8\frac{\psi}{\psi_a} - 3\frac{\psi^2}{\psi_a^2}\right) \tag{3.3.14}$$

with $\psi_a = \frac{1}{2}B_0 a^2$.

3.4 Hamiltonian of the DED field

For the mapping of the perturbed system, we need the Hamiltonian of the total magnetic field, including the perturbation field of the DED coils. In the previous section we specified the main helical field $\vec{B}_g$, especially the safety factor, which describes the poloidal field of the plasma current, and its Hamiltonian. Now we include the perturbation field, described by the magnetic scalar potential

$$\phi(r,\theta,\varphi) = \sum_{m=-\infty}^{\infty} \phi_m I_m(\frac{n_0 r}{R_0}) \cos(m\theta - n_0\varphi + \omega t), \qquad (3.4.1)$$

according to Eq. (3.2.44). The magnetic field of the DED is then given by

$$\vec{B}_s = -\nabla \phi(r,\theta,\varphi) \qquad (3.4.2)$$

with its components

$$B_r = -\frac{\partial \phi}{\partial r} = -\sum_{m=-\infty}^{\infty} \phi_m \left[\frac{n_0}{R_0} I_{m-1}(\frac{n_0 r}{R_0}) - \frac{m}{r} I_m(\frac{n_0 r}{R_0}) \right] \cos(m\theta - n_0\varphi + \omega t) \qquad (3.4.3)$$

$$B_\theta = -\frac{1}{r}\frac{\partial \phi}{\partial \theta} = \sum_{m=-\infty}^{\infty} \phi_m \frac{m}{r} I_m(\frac{n_0 r}{R_0}) \sin(m\theta - n_0\varphi + \omega t) \qquad (3.4.4)$$

$$B_\varphi = -\frac{1}{R_0}\frac{\partial \phi}{\partial \varphi} = -\sum_{m=-\infty}^{\infty} \phi_m \frac{n_0}{R_0} I_m(\frac{n_0 r}{R_0}) \sin(m\theta - n_0\varphi + \omega t). \qquad (3.4.5)$$

We use again the Clebsch form with the ansatz

$$\vec{B} = \nabla\psi \times \nabla\theta^* - \nabla H \times \nabla\varphi = \begin{pmatrix} \frac{\partial \psi}{\partial r} \\ \frac{1}{r}\frac{\partial \psi}{\partial \theta} \\ \frac{\partial \psi}{\partial z} \end{pmatrix} \times \begin{pmatrix} \frac{\partial \theta^*}{\partial r} \\ \frac{1}{r}\frac{\partial \theta^*}{\partial \theta} \\ \frac{\partial \theta^*}{\partial z} \end{pmatrix} - \begin{pmatrix} \frac{\partial H}{\partial r} \\ \frac{1}{r}\frac{\partial H}{\partial \theta} \\ \frac{\partial H}{\partial z} \end{pmatrix} \times \begin{pmatrix} 0 \\ 0 \\ \frac{1}{R_0} \end{pmatrix}, \qquad (3.4.6)$$

but now for the total magnetic field $\vec{B} = \vec{B}_g + \vec{B}_s$

$$\vec{B} = \begin{pmatrix} -\sum_{m=-\infty}^{\infty} \phi_m \frac{n_0}{R_0} \left[I_{m-1}(x) - \frac{m}{x} I_m(x) \right] \cos(m\theta - n_0\varphi + \omega t) \\ \frac{\mu_0 I_p(r)}{2\pi r} + \sum_{m=-\infty}^{\infty} \phi_m \frac{m}{r} I_m(x) \sin(m\theta - n_0\varphi + \omega t) \\ B_0 - \sum_{m=-\infty}^{\infty} \phi_m \frac{n_0}{R_0} I_m(x) \sin(m\theta - n_0\varphi + \omega t) \end{pmatrix}, \qquad (3.4.7)$$

where we use the shortcut $x = \frac{n_0 r}{R_0}$. Because the DED field $\vec{B}_s$ is a small perturbation to the main field $\vec{B}_g$, we can solve Eq. (3.4.6) in each order of the perturbation separately. Therefore,

3.4 Hamiltonian of the DED field

we set
$$H = H_0 + H_1 \quad \text{and} \quad \theta^* = \theta_0 + \theta_1 , \tag{3.4.8}$$

while ψ remains unperturbed. In the zeroth order we get the already known results from Sec. 3.3

$$\Rightarrow \quad \psi = \frac{1}{2}B_0 r^2, \quad \theta_0 = \theta \quad \text{and} \quad H_0(r) = \frac{\mu_0 R_0}{2\pi} \int \frac{I_p(r)}{r} dr = \int \frac{d\psi}{q(\psi)} . \tag{3.4.9}$$

Including the perturbation, we get from the radial component

$$\sum_{m=-\infty}^{\infty} \phi_m n_0 r \left[I_{m-1}(x) - \frac{m}{x} I_m(x) \right] \cos(m\theta - n_0\varphi + \omega t) = \frac{\partial H}{\partial \theta} \tag{3.4.10}$$

$$\Rightarrow \quad H = \sum_{m=-\infty}^{\infty} \phi_m R_0 \left[\frac{x}{m} I_{m-1}(x) - I_m(x) \right] \sin(m\theta - n_0\varphi + \omega t) + H_0(r) . \tag{3.4.11}$$

For the poloidal component we need the derivative of H with respect to r

$$\frac{\partial H}{\partial r} = \sum_{m=-\infty}^{\infty} \phi_m n_0 \frac{\partial}{\partial x} \left[\frac{x}{m} I_{m-1}(x) - I_m(x) \right] \sin(m\theta - n_0\varphi + \omega t) + \frac{\partial H_0}{\partial r} , \tag{3.4.12}$$

while the derivative of the Bessel functions reads

$$\frac{\partial}{\partial x} \left[\frac{x}{m} I_{m-1}(x) - I_m(x) \right] = \left(\frac{x}{m} + \frac{m}{x} \right) I_m(x) . \tag{3.4.13}$$

From the poloidal component

$$\frac{R_0 \mu_0 I_p(r)}{2\pi r} + \sum_{m=-\infty}^{\infty} \phi_m n_0 \frac{m}{x} I_m(x) \sin(m\theta - n_0\varphi + \omega t) = -B_0 r \frac{\partial \theta^*}{\partial \varphi} + \frac{\partial H}{\partial r} \tag{3.4.14}$$

we get, using Eq. (3.4.12) with Eq. (3.4.13),

$$\sum_{m=-\infty}^{\infty} \phi_m n_0 \frac{m}{x} I_m(x) \sin(m\theta - n_0\varphi + \omega t) \tag{3.4.15}$$

$$= -B_0 r \frac{\partial \theta^*}{\partial \varphi} + \sum_{m=-\infty}^{\infty} \phi_m n_0 \left(\frac{x}{m} + \frac{m}{x} \right) I_m(x) \sin(m\theta - n_0\varphi + \omega t) \tag{3.4.16}$$

$$\Leftrightarrow \quad B_0 r \frac{\partial \theta^*}{\partial \varphi} = \sum_{m=-\infty}^{\infty} \phi_m n_0 \frac{x}{m} I_m(x) \sin(m\theta - n_0\varphi + \omega t) \tag{3.4.17}$$

$$\Rightarrow \quad \theta^* = \theta + \sum_{m=-\infty}^{\infty} \phi_m \frac{n_0}{m R_0 B_0} I_m(x) \cos(m\theta - n_0\varphi + \omega t) . \tag{3.4.18}$$

Chapter 3. Cylindrical model for magnetic field lines in TEXTOR-DED

Inserting the results (3.4.9) and (3.4.18) into the z-component, we find

$$rB_0 - \sum_{m=-\infty}^{\infty} \phi_m x I_m(x) \sin(m\theta - n_0\varphi + \omega t) = \frac{\partial \psi}{\partial r}\frac{\partial \theta^*}{\partial \theta} - \frac{\partial \psi}{\partial \theta}\frac{\partial \theta^*}{\partial r}$$

$$= B_0 r \left(1 - \sum_{m=-\infty}^{\infty} \phi_m \frac{n_0}{R_0 B_0} I_m(x) \sin(m\theta - n_0\varphi + \omega t)\right)$$

$$\Leftrightarrow 0 = 0,$$

which means that the z-component is fulfilled. Finally, we obtain for the Hamiltonian description $H = H(\psi, \theta^*, \varphi)$ of the cylindrical DED model

$$H = \int \frac{d\psi}{q(\psi)} + \sum_{m=-\infty}^{\infty} \phi_m R_0 \left[\frac{x}{m} I_{m-1}(x) - I_m(x)\right] \sin(m\theta - n_0\varphi + \omega t) \qquad (3.4.19)$$

$$\theta^* = \theta + \sum_{m=-\infty}^{\infty} \phi_m \frac{n_0}{mR_0 B_0} I_m(x) \cos(m\theta - n_0\varphi + \omega t) \qquad (3.4.20)$$

$$\psi = \frac{1}{2} B_0 r^2, \qquad (3.4.21)$$

with the q-profile (3.3.11) and $x = \frac{n_0 r}{R_0}$

3.5 The DED map

To derive the mapping procedure for the Poincaré plot of the cylindrical DED model from the Hamiltonian description above, we have to calculate the generating function S, according to Sec. 2.2. We want to derive the generating function of the map up to the first order of the perturbation. Only the perturbation part of the Hamiltonian depends on θ^*, so that θ_1 inserted in H_1 would be of second order of the perturbation. For this reason, we can neglect the perturbation part θ_1 of θ^* and use $\theta^* = \theta$ in the following. We also approximate the Bessel functions I and K according to Eq. (3.2.42). For the argument $x = \frac{n_0 r}{R_0} < 1.22$ is valid, using the parameters of our model. This means, $x \ll m$ is valid for all $|m| \geq 2$. In ϕ_m only the mode numbers m around the poloidal main mode number $m_0 (= 20)$ and its uneven multiples are the relevant modes, because of the Fourier spectrum parameter $g_m^{(k)}$, see Eq. (3.1.15). Therefore, the errors due to the approximation are negligibly small. Then the simplified Hamiltonian for the DED model reads

$$H(\psi, \theta, \varphi) = \int \frac{d\psi}{q(\psi)} + \sum_{m=-\infty}^{\infty} G_m(\psi) \sin(m\theta - n_0\varphi + \omega t) \qquad (3.5.1)$$

3.5 The DED map

with

$$G_m(\psi) = (-1)^{m-n_0+1}\frac{1}{2}\mu_0 J_0 R_0 r_c \cos\alpha \frac{\sin((m-m_0)\theta_c)}{m(m-m_0)\pi}\left(\frac{\psi}{\psi_c}\right)^{m/2} \quad (3.5.2)$$

and $\psi_c = \frac{1}{2}B_0 r_c^2$. To calculate the first order generating function S, we have to integrate the perturbation part of the Hamiltonian along the unperturbed trajectory

$$\psi = \text{const}, \quad \theta(\varphi) = \theta_0 + \Omega(\psi)(\varphi - \varphi_0), \quad (3.5.3)$$

which is given by the winding number Ω, defined by Eq. (3.3.12), over the "time" φ. We obtain

$$\begin{aligned}
S &= -\sum_m G_m(\psi)\int_{\varphi_0}^{\varphi}\sin(m\theta(\varphi') - n_0\varphi' + \omega t)\,d\varphi' \\
&= -\sum_m G_m(\psi)\int_{\varphi_0}^{\varphi}\sin(m\theta + m\Omega(\varphi' - \varphi) - n_0\varphi' + \omega t)\,d\varphi' \\
&= \sum_m G_m(\psi)\frac{1}{m\Omega - n_0}[\cos(m\theta - n_0\varphi + \omega t) - \cos(m\theta - m\Omega(\varphi - \varphi_0) - n_0\varphi_0 + \omega t)] \\
&= -\sum_m 2G_m(\psi)\frac{\sin(\frac{1}{2}(\varphi - \varphi_0)(m\Omega - n_0))}{m\Omega - n_0} \\
&\quad \times \sin(m\theta - \frac{1}{2}m\Omega(\varphi - \varphi_0) - \frac{1}{2}n_0(\varphi + \varphi_0) + \omega t)\,.
\end{aligned}$$

Introducing the following shortcuts

$$z := m\Omega - n_0\,,\quad y := m\theta - \frac{1}{2}m\Omega(\varphi - \varphi_0) - \frac{1}{2}n_0(\varphi + \varphi_0) + \omega t\,,\quad k := \frac{1}{2}(\varphi - \varphi_0) \quad (3.5.4)$$

and

$$h_1(z) := \frac{\sin(kz)}{z} \to k \quad \text{for } z \to 0\,, \quad (3.5.5)$$

the first order generating function reads

$$S = -\sum_m 2G_m(\psi)h_1(z)\sin(y)\,. \quad (3.5.6)$$

In Appendix C we present another form of the generating function and its derivatives, which correspond to the general form shown in Sec. 2.2. The form shown in the appendix, which is also used in literature [5], is totally equivalent to the one above.

We have to calculate the derivatives of S with respect to ψ and θ, respectively, to perform the iteration procedure of the map. We further need the second, mixed derivative for the numerics to use Newton's method for solving the implicit equations of the map. For the derivative of S

Chapter 3. Cylindrical model for magnetic field lines in TEXTOR-DED

with respect to ψ, we need

$$\frac{\partial}{\partial \psi} h_1(z) = m\Omega'(\psi) \frac{\partial h_1}{\partial z}, \tag{3.5.7}$$

$$\frac{\partial h_1}{\partial z} =: h_1'(z) = \frac{1}{z^2}(k\cos(kz)z - \sin(kz)) = \frac{k\cos(kz) - h_1(z)}{z} \to 0 \quad \text{for } z \to 0 \tag{3.5.8}$$

and

$$G_m'(\psi) = \frac{m}{2} \frac{G_m(\psi)}{\psi}, \tag{3.5.9}$$

to obtain

$$\begin{aligned}
\frac{\partial S}{\partial \psi} &= -2 \sum_m \left\{ \left[G_m'(\psi) h_1(z) + m\Omega' G_m(\psi) h_1'(z) \right] \sin(y) - mk\Omega' G_m(\psi) h_1(z) \cos(y) \right\} \\
&= \sum_m m G_m(\psi) \left\{ \left[-\frac{\sin(y)}{\psi} + 2k\Omega' \cos(y) \right] h_1(z) - 2\Omega' h_1'(z) \sin(y) \right\}.
\end{aligned} \tag{3.5.10}$$

The derivative with respect to θ reads

$$\frac{\partial S}{\partial \theta} = -2 \sum_m m G_m(\psi) h_1(z) \cos(y), \tag{3.5.11}$$

and for the second, mixed one, we get

$$\frac{\partial^2 S}{\partial \psi \partial \theta} = \frac{\partial^2 S}{\partial \theta \partial \psi} = \sum_m m^2 G_m(\psi) \left\{ \left[-\frac{\cos(y)}{\psi} - 2k\Omega' \sin(y) \right] h_1(z) - 2\Omega' h_1'(z) \cos(y) \right\}. \tag{3.5.12}$$

Now we can calculate the map by the iteration procedure

$$\xi_k = \psi_k - \frac{\partial S_k}{\partial \theta_k}, \qquad \vartheta_k = \theta_k + \frac{\partial S_k}{\partial \xi_k} \tag{3.5.13}$$

$$\vartheta_{k+1} = \vartheta_k + \Omega(\xi_k)(\varphi_{k+1} - \varphi_k) \tag{3.5.14}$$

$$\theta_{k+1} = \vartheta_{k+1} - \frac{\partial S_{k+1}}{\partial \xi_k}, \qquad \psi_{k+1} = \xi_k + \frac{\partial S_{k+1}}{\partial \theta_{k+1}} \tag{3.5.15}$$

with $S_k = S(\xi_k, \theta_k, \varphi_k)$ and $S_{k+1} = S(\xi_k, \theta_{k+1}, \varphi_{k+1})$. This is the so called DED map in its symmetric form. The first equations of (3.5.13) and (3.5.15), respectively, are implicit equations, which have to be solved by Newton's method, using the mixed derivative (3.5.12) of S. A huge problem is that the solutions of these implicit equations are not unique. There are many different solutions, depending on the initial parameters of the Newton method. This problem can only be solved by performing the iteration with a small step size, like $d\varphi = \varphi_{k+1} - \varphi_k = 2\pi/16$, which is used for the numerical results presented here. Compared to the tokamap, where one full toroidal turn is performed at each step of the map, the DED map takes much more computational time. But, compared to the direct numerical integration of

3.5 The DED map

the Hamiltonian equations of motion, where over a thousand steps are needed for one toroidal turn, the mapping is much faster, even if one would use 32 or 64 steps.

In contrast to the tokamap, the generating function of the DED map depends explicitly on the free parameter φ_0, which specifies the reference time within the finite time interval of one mapping step, according to Sec. 2.2. So, we have to choose φ_0 reasonably. Remember, the choice of φ_0 specifies, which type of map will be created. One could choose $\varphi_0 = \varphi_k$, setting the reference time at the beginning of the time interval $d\varphi$. Then Eq. (3.5.13) would vanish, because of $S_k \equiv 0$. Choosing the reference time at the end of $d\varphi$, meaning $\varphi_0 = \varphi_{k+1}$, would lead to $S_{k+1} \equiv 0$, causing Eq. (3.5.15) to vanish. These two cases would produce twist maps. Due to the lack of symmetry, it is extremely difficult to invert a twist map properly. To choose the reference time in the middle of the interval $d\varphi$, is the most convenient choice, which means $\varphi_0 = \frac{1}{2}(\varphi_{k+1} + \varphi_k)$. This produces the symmetric map, which has several advantages compared to the twist map. For example, the symmetric map preserves the reverse time symmetry of the system and can therefore be inverted easily. Its structure is totally invariant to time inversion. The symmetric map is also more precise than the twist map, according to [9].

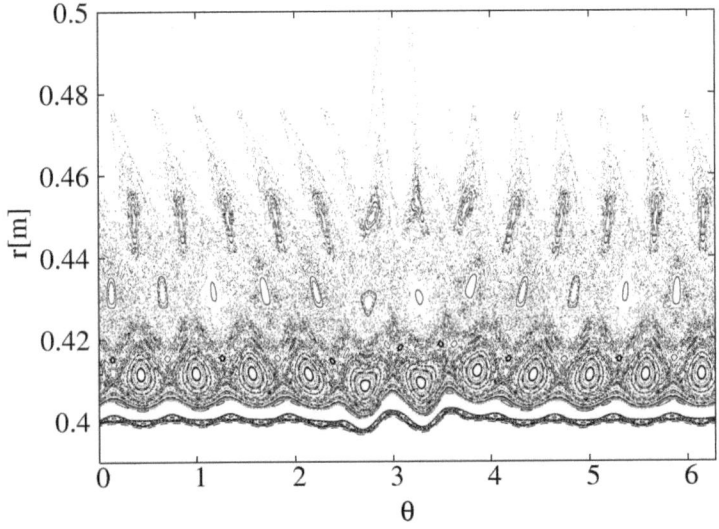

Figure 3.4: Symmetric DED map with $I_0 = 10$ kA, $B_0 = 2.2$ T, q-profile "a" with $I_p = 390$ kA in the cylindrical model.

Figure 3.4 shows the DED map for the cylindrical model, using the TEXTOR parameters, $R_0 = 1.75$ m and $r_c = 0.5325$ m. The perturbation of the DED is given by the current of the DED coils of $I_0 = 10$ kA. We are using the q-profile "a", specified in Sec. 3.3, with a total

Chapter 3. Cylindrical model for magnetic field lines in TEXTOR-DED

plasma current of $I_p = 390$ kA and the main magnetic field $B_0 = 2.2$ T. Every point in this Poincaré plot is an intersection point of a magnetic field line with the $\varphi = 0 \mod 2\pi$ plane. As one can see, the DED map is an open chaotic system like the revtokamap, whereas the wall is located at the radial position $r_w = 0.5$ m. The finger-like structures are typical for the DED map and characterizes the so called laminar zone, where the field lines have very short connection lengthes, the length of the field line from wall to wall inside the cylinder. The region next to the laminar zone is the ergodic zone, which is dominated by remaining island chains surrounded by a chaotic sea. In Fig. 3.4 we have three island chains, corresponding to the rational resonant q-surfaces $q = 11/4$ at $r = 0.41$ m, $q = 12/4 = 3$ at $r = 0.43$ m and $q = 13/4$ at $r = 0.45$ m. Because we have neglected all higher poloidal modes in Eq. (3.2.44), only the $n = n_0 = 4$ poloidal mode is resonant in the DED map. The first intact KAM surface is located below the 11/4 island chain and separates the inside of the cylinder from the wall. All interactions, like transport of magnetic field lines, take place in the ergodic and laminar zones. In the following, we focus on these regions only.

3.6 Characterization of the DED system by its statistical properties

To get a first insight into the structures and dynamics, like transport mechanisms, we have to classify the chaotic system by its statistical properties. Therefore, we investigate how the magnetic field lines of the ergodic and laminar zone are connected to the wall. One interesting question is the diffusion mechanism of the field lines. To determine the diffusion, we use the mean square displacement (MSD)

$$\sigma_\psi(n) = \langle (\psi_n - \langle \psi_n \rangle)^2 \rangle \ , \tag{3.6.1}$$

already introduced in Sec. 2.4. The average $\langle \ldots \rangle$ is taken over the initial points. Then the diffusion coefficient is given by the derivative with respect to time

$$D = \frac{1}{2} \frac{d\sigma_\psi(n)}{dn} \ , \tag{3.6.2}$$

whereas n is the number of iterations, dealing as time in this case.

Figures 3.5 and 3.6 show the MSD of the DED map for short and long time behavior, respectively, for different perturbation currents I_0. We have taken 20000 initial points at various angle positions within the chaotic sea at the fixed radial position $r = 0.412$ m, using the same parameter configuration as used for Fig. 3.4. The short time behavior of the MSD, Fig. 3.5, shows an increase of the MSD with a gradient less than linear. This indicates subdiffusive transport behavior. As one can also see from Fig. 3.5, the transport increases with increasing perturbation, which is totally understandable, because the increasing perturbation destroys

3.6 Characterization of the DED system by its statistical properties

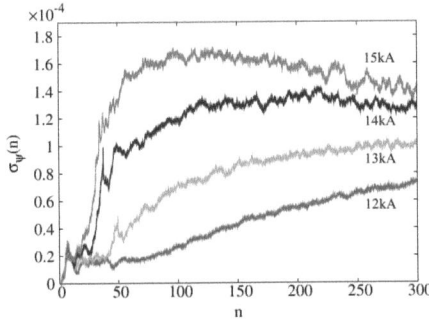

Figure 3.5: MSD of the flux ψ for various perturbation currents I_0

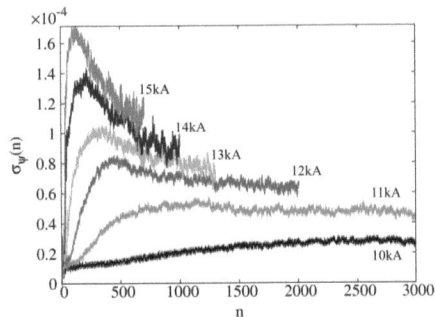

Figure 3.6: Same as Fig. 3.5 for long iteration time

more and more the remaining island chains, which constrict the transport enormously. If the perturbation is increased so much that all islands and structures have vanished, the transport would be diffusive, as we have seen from the large perturbation analysis of the tokamap in Sec. 2.4. So, the islands are playing a great role in the transport analysis, which can also be seen in Fig. 3.6. This figure shows, that the MSD drops after a certain time. Usually, the MSD should not decrease, but in our case we are loosing field lines at the wall, so only the field lines, which are staying in the system for extremely long time, remain. These are the field lines, which are sticking around the islands, so the MSD drops to the characteristical width of the island chains. This width decreases for smaller perturbations, because the islands are smaller and the area around the island is more regular. There are, for example, more substructures so that the field lines which are sticking to the islands are bound much closer to the islands than for larger perturbations. The time, when the drop occurs, is smaller for larger perturbations. This is also a clear result. Due to the increased transport for larger perturbations, more field lines are reaching the wall in shorter time.

The diffusion coefficient provides us with the information, how the field lines are transported to the wall collectively. But within the chaotic layer the field lines are also moving away from each other, because neighboring field lines are always moving in completely different ways and directions. It is important to see, how the field lines are separating from each other. By definition of chaotic motion, the field lines are separating exponentially, which means for their distance s_N after N iterations

$$s_N = s_0\, e^{\chi \cdot N} \qquad (3.6.3)$$

with the distance s_0 at the beginning. The characteristical quantity for this behavior is the

Chapter 3. Cylindrical model for magnetic field lines in TEXTOR-DED

Lyapunov exponent χ [4], which is then defined by

$$\chi = \lim_{N \to \infty} \frac{1}{N} \ln \frac{s_N}{s_0}. \tag{3.6.4}$$

To calculate the Lyapunov exponent, we have to derive the distance of the field lines from the mapping. Therefore, we assume the infinitesimal vector

$$d\vec{V}_k = \begin{pmatrix} d\psi_k \\ d\theta_k \end{pmatrix}, \tag{3.6.5}$$

which describes the distance between two neighboring field lines at the k-th iteration of the map. The development of this vector under the map is given by the equation

$$d\vec{V}_{k+1} = M_k d\vec{V}_k, \tag{3.6.6}$$

where M_k is the Jacobi matrix

$$M_k = \begin{pmatrix} \frac{\partial \psi_{k+1}}{\partial \psi_k} & \frac{\partial \psi_{k+1}}{\partial \theta_k} \\ \frac{\partial \theta_{k+1}}{\partial \psi_k} & \frac{\partial \theta_{k+1}}{\partial \theta_k} \end{pmatrix} \tag{3.6.7}$$

of the map. For the symmetric map (2.2.11)-(2.2.13) this matrix can be written as a product of three matrices

$$M_k = M_k^{(3)} M_k^{(2)} M_k^{(1)} \tag{3.6.8}$$

with

$$M_k^{(3)} = \begin{pmatrix} \frac{\partial \psi_{k+1}}{\partial \xi_{k+1}} & \frac{\partial \psi_{k+1}}{\partial \vartheta_{k+1}} \\ \frac{\partial \theta_{k+1}}{\partial \xi_{k+1}} & \frac{\partial \theta_{k+1}}{\partial \vartheta_{k+1}} \end{pmatrix} \quad M_k^{(2)} = \begin{pmatrix} 1 & 0 \\ \Omega'(\xi_k)(\varphi_{k+1} - \varphi_k) & 1 \end{pmatrix} \quad M_k^{(1)} = \begin{pmatrix} \frac{\partial \xi_k}{\partial \psi_k} & \frac{\partial \xi_k}{\partial \theta_k} \\ \frac{\partial \vartheta_k}{\partial \psi_k} & \frac{\partial \vartheta_k}{\partial \theta_k} \end{pmatrix}. \tag{3.6.9}$$

The distance of the field lines after k iterations is then given by the length of the vector, which means

$$s_k^2 = d\psi_k^2 + d\theta_k^2 = d\vec{V}_k^T d\vec{V}_k = d\vec{V}_{k-1}^T M_k^T M_k d\vec{V}_{k-1} = d\vec{V}_0^T M_1^T \cdots M_k^T M_k \cdots M_1 d\vec{V}_0. \tag{3.6.10}$$

This distance can be approximated by using the eigenvalues of the matrix M_k. As a two dimensional matrix, M_k has two eigenvalues $\lambda_{k_{1/2}}$, where we can generally assume $|\lambda_{k_1}| \geq |\lambda_{k_2}|$. Therefore, we obtain for the square of the distance

$$s_k^2 \leq s_0^2 \prod_{k=1}^{N} |\lambda_{k_1}|^2 \tag{3.6.11}$$

3.6 Characterization of the DED system by its statistical properties

so that we get for the maximum Lyapunov exponent

$$\chi = \lim_{N \to \infty} \frac{1}{N} \sum_{k=1}^{N} \ln(|\lambda_{k_1}|) \,. \tag{3.6.12}$$

Now we need the eigenvalues of the matrix M_k. We introduce the shortcuts

$$\frac{\partial^2 S_k}{\partial \xi_k^2} = S_{\xi\xi}^{(k)} \qquad \frac{\partial^2 S_k}{\partial \theta_k^2} = S_{\theta\theta}^{(k)} \qquad \frac{\partial^2 S_k}{\partial \xi_k \partial \theta_k} = S_{\xi\theta}^{(k)} \tag{3.6.13}$$

and similar for S_{k+1}. To calculate the entries of the matrix $M_k^{(1)}$, we use the iteration equations and obtain

$$\frac{\partial \xi_k}{\partial \psi_k} = 1 - S_{\xi\theta}^{(k)} \frac{\partial \xi_k}{\partial \psi_k} \Rightarrow \frac{\partial \xi_k}{\partial \psi_k} = \frac{1}{1 + S_{\xi\theta}^{(k)}} =: f_k \,,$$

$$\frac{\partial \xi_k}{\partial \theta_k} = -S_{\theta\theta}^{(k)} - S_{\xi\theta}^{(k)} \frac{\partial \xi_k}{\partial \theta_k} \Rightarrow \frac{\partial \xi_k}{\partial \theta_k} = -f_k S_{\theta\theta}^{(k)} \,,$$

$$\frac{\partial \vartheta_k}{\partial \psi_k} = f_k S_{\xi\xi}^{(k)} \,,$$

$$\frac{\partial \vartheta_k}{\partial \theta_k} = 1 + S_{\xi\theta}^{(k)} + S_{\xi\xi}^{(k)} \frac{\partial \xi_k}{\partial \theta_k} = \frac{1}{f_k} - f_k S_{\xi\xi}^{(k)} S_{\theta\theta}^{(k)} \,,$$

and similar for $M_k^{(3)}$ with $\xi_{k+1} = \xi_k$

$$\frac{\partial \theta_{k+1}}{\partial \vartheta_{k+1}} = 1 - S_{\xi\theta}^{(k+1)} \frac{\partial \theta_{k+1}}{\partial \vartheta_{k+1}} \Rightarrow \frac{\partial \theta_{k+1}}{\partial \vartheta_{k+1}} = \frac{1}{1 + S_{\xi\theta}^{(k+1)}} =: f_{k+1} \,,$$

$$\frac{\partial \theta_{k+1}}{\partial \xi_k} = -S_{\xi\xi}^{(k+1)} - S_{\xi\theta}^{(k+1)} \frac{\partial \theta_{k+1}}{\partial \xi_k} \Rightarrow \frac{\partial \theta_{k+1}}{\partial \xi_k} = -f_{k+1} S_{\xi\xi}^{(k+1)} \,,$$

$$\frac{\partial \psi_{k+1}}{\partial \vartheta_{k+1}} = f_{k+1} S_{\theta\theta}^{(k+1)} \,,$$

$$\frac{\partial \psi_{k+1}}{\partial \xi_k} = 1 + S_{\xi\theta}^{(k+1)} + S_{\theta\theta}^{(k+1)} \frac{\partial \theta_{k+1}}{\partial \xi_k} \Rightarrow \frac{\partial \psi_{k+1}}{\partial \xi_k} = \frac{1}{f_{k+1}} - f_{k+1} S_{\xi\xi}^{(k+1)} S_{\theta\theta}^{(k+1)} \,.$$

Combining all results, we obtain for M_k

$$M_k = f_k f_{k+1} \begin{pmatrix} \frac{1}{f_{k+1}^2} + g_k S_{\theta\theta}^{(k+1)} & -\frac{1}{f_{k+1}^2} S_{\theta\theta}^{(k)} + \frac{1}{f_k^2} S_{\theta\theta}^{(k+1)} - g_k S_{\theta\theta}^{(k)} S_{\theta\theta}^{(k+1)} \\ g_k & \frac{1}{f_k^2} - g_k S_{\theta\theta}^{(k)} \end{pmatrix} \tag{3.6.14}$$

with

$$g_k := \Omega'(\xi_k)(\varphi_{k+1} - \varphi_k) + S_{\xi\xi}^{(k)} - S_{\xi\xi}^{(k+1)} \,. \tag{3.6.15}$$

It can easily be seen that $\det(M_k) = 1$ holds, because of the symplectic probabilities of the

55

map. Then we can obtain the eigenvalues by the zero points

$$\lambda_{k_{1/2}} = \frac{\text{Tr}(M_k)}{2} \pm \sqrt{\frac{\text{Tr}(M_k)^2}{4} - \det(M_k)} \qquad (3.6.16)$$

of the characteristic equation, whereas the trace is given by

$$\text{Tr}(M_k) = f_k f_{k+1} \left[\frac{1}{f_{k+1}^2} + \frac{1}{f_k^2} + g_k(S_{\theta\theta}^{(k+1)} - S_{\theta\theta}^{(k)}) \right] . \qquad (3.6.17)$$

To apply this calculation method to the DED map, it is necessary to calculate the second derivatives of the generating function S, Eq. (3.5.6), with respect to ψ and θ, respectively. They read

$$\begin{aligned}
\frac{\partial^2 S}{\partial \psi^2} &= \sum_m m G_m(\psi) \Bigg\{ \left[\left(-\frac{m-2}{2\psi^2} - 2k^2 m \Omega'^2 \right) \sin(y) + \left(\frac{2mk}{\psi} \Omega' + 2k\Omega'' \right) \cos(y) \right] h_1(z) \\
&\quad + \left[\left(-\frac{2m}{\psi} \Omega' - 2\Omega'' \right) \sin(y) + 4mk\Omega'^2 \cos(y) \right] h_1'(z) - 2m\Omega'^2 \sin(y) h_1''(z) \Bigg\} \quad (3.6.18) \\
\frac{\partial^2 S}{\partial \theta^2} &= 2 \sum_m m^2 G_m(\psi) h_1(z) \sin(y) , \qquad (3.6.19)
\end{aligned}$$

with the second derivative of $h_1(z)$ with respect to z

$$h_1''(z) = -k^2 h_1(z) - 2\frac{h_1'(z)}{z} \to -\frac{1}{3}k^3 \quad \text{for } z \to 0 . \qquad (3.6.20)$$

Figure 3.7 shows the dependency of the Lyapunov exponent on the poloidal angle θ at the fixed radial position $r = 0.425$ m for the perturbation current $I_0 = 10$ kA. As one can see, the Lyapunov exponent fluctuates stochastically around a constant mean value $\bar{\chi}(r)$ and the structure is fractal. At other radial positions the Lyapunov exponent shows the same behavior, so there is no particular dependency on the poloidal angle, except inside an island, where the Lyapunov exponent is equal to zero, because the orbits inside an island are stable and neighboring field lines there will always stay together. The mean Lyapunov exponent depends only on the radial position.

In order to obtain a quantitative measure of the exponential divergence of field lines along the radius, we introduce the Kolmogorov length

$$L_k(r) = \frac{2\pi R_0}{\bar{\chi}(r)} . \qquad (3.6.21)$$

It uses the reciprocal value of the mean Lyapunov exponent. The Kolmogorov length characterizes a correlation length of field lines along the height of the cylinder $z = 2\pi R_0$. From the

3.6 Characterization of the DED system by its statistical properties

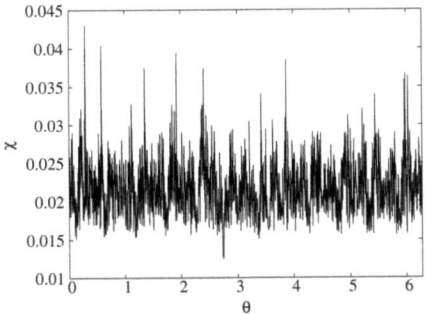

Figure 3.7: Lyapunov exponent depending on the poloidal angle θ at a fixed radial position of $r = 0.425$ m for $I_0 = 10$ kA.

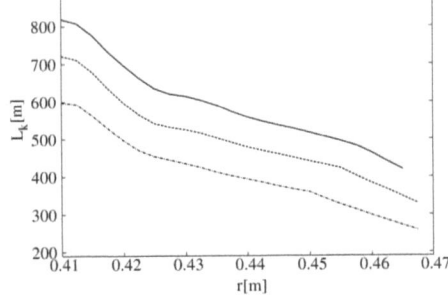

Figure 3.8: Kolmogorov length depending on the radius for three different perturbation currents: $I_0 = 10$ kA solid line, $I_0 = 12$ kA dashed line and $I_0 = 15$ kA dot-dashed line.

definition (3.6.4) of the Lyapunov exponent we get

$$s(l) = s(0)e^{l/L_k}, \qquad (3.6.22)$$

whereas s is the distance between the field lines and l is their length. Figure 3.8 shows the dependency of the Kolmogorov length on the radius for different perturbation currents. The Kolmogorov length decreases with increasing perturbation, which is a clear result, because the greater the perturbation, the faster the field lines diverge. More interesting is that we can distinguish three different radial zones, where the curves show significant changes. Compared to Fig. 3.4, we can identify the certain zones. The first one is up to $r = 0.425$ m for $I_0 = 10$ kA, where the Kolmogorov length drops fast. This is the area around the 11/4 island chain, where the most remaining structures are located, due to the proximity to the regular inside, where the Kolmogorov length is infinite. So, the Kolmogorov length has to drop fast close to this regular inside. The next zone is up to $r = 0.455$ m. This is clearly the ergodic zone. Within this zone the Kolmogorov length decreases slowly, because the field lines in the ergodic zone are mainly chaotic, but still bound to the islands. The third zone, beyond $r = 0.46$ m, is the laminar zone, where the field lines are going to the wall very fast and therefore the Kolmogorov length drops very fast. One can also see that with increasing perturbation all zones are growing while shifting inwards. The Kolmogorov length clearly reflects the different zones and their properties.

3.7 Topology of the stochastic edge region, analyzed by the stable and unstable manifolds

For the tokamap we already discussed the importance of the stable and unstable manifolds, their important role in the creation of chaotic layers, and their connection to chaotic transport, see Secs. 2.7 and 2.9. These results are fundamental and also valid for the DED map. But the DED map is an open chaotic system and as we shall see, the manifolds have a large influence on the structures of the Poincaré section and the wall patterns. Therefore, we calculate the unstable manifolds of hyperbolic periodic points of the DED map, using the algorithm described in Sec. 2.5. To calculate the stable and unstable manifolds, we have to find the hyperbolic points first. Using a two-dimensional Newton method or the minimization method, described in Sec. 2.6, the hyperbolic points of the 12/4 island chain are calculated and shown as crosses in Fig. 3.9.

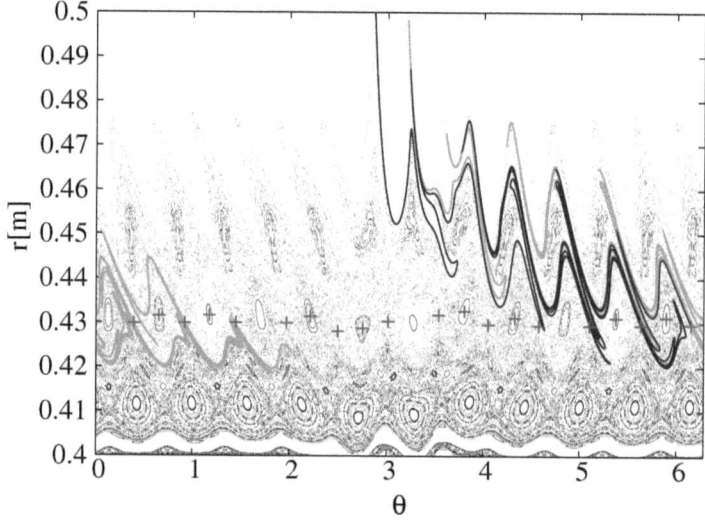

Figure 3.9: Symmetric DED map with $I_0 = 10$ kA, similar to Fig. 3.4, with period 3 periodic points, crosses, left sided, black line, and right sided, grey line, unstable manifold of the hyperbolic point at $\theta \approx 6.15$.

Figure 3.9 also shows the left and right sided unstable manifold of a period 3 hyperbolic point of the DED map with the perturbation current $I_0 = 10$ kA. In the already known manner, the manifold goes around the islands and starts oscillating when it closes the next hyperbolic point. The amplitude of these oscillations increases very fast so that the loops are also going

3.7 Topology of the stochastic edge region, analyzed by the stable and unstable manifolds

around the islands and so on. This leads to the sticking of the field lines around the islands, already observed in the statistical properties of the previous section. In Fig. 3.9 the sticking can be seen very well, especially on the right sided part of the unstable manifold, the grey line. Due to the overlap of the chaotic layers of the several island chains, caused by the intersection of their stable and unstable manifolds, as shown in Figs. 2.8 and 2.9, the field lines are transported from one island chain to the next one along the stable and unstable manifolds. In Fig. 3.9 the unstable manifold of the period 3 hyperbolic point tends towards the wall and for this, it has to pass the next, 13/4 island chain. The unstable manifold merges with the unstable manifold of the 13/4 hyperbolic points. They do not intersect with each other, but they converge against each other and become infinitesimally close. Beyond the 13/4 island chain there are no further islands, so the manifold hits the wall. Now we can draw the most important conclusions, clearly seen in Fig. 3.9. The finger-like structures are formed by the unstable manifolds. Therefore, the wall pattern, which is an imprint of the fingers at the wall, is fundamentally dominated by the unstable manifolds. This means that the heat and particle deposition at the wall as well as their transport within the plasma is dominated by the unstable manifolds. The structures and the developing of the unstable manifolds close to the wall are dominated by the last remaining island chain, because all manifolds of island chains below have to pass the last island chain. Beyond the last island chain there are no more hyperbolic points to create further oscillations. So, the manifold goes to the wall straightly. This causes the field lines to hit the wall very fast after transported into this region, which, as already mentioned above, characterizes the laminar zone. The last remaining resonance in front of the wall has the most influence on the wall pattern and the heat and particle deposition, if we assume that the thermal particles follow the field lines.

To further analyze the laminar zone, laminar plots are a reasonable tool, which we also used on the revtokamap in Sec. 2.9. Remember, a laminar plot is a colored contour plot. Areas of field lines, whose number of toroidal rotations are the same until they connect the wall through this area, are colored in the same way. The laminar plot then gives information about the connection lengthes of the field lines.

Figure 3.10 shows a laminar plot of the two main fingers around $\theta = \pi$ of the DED map for the perturbation current $I_0 = 10$ kA. The unstable manifold, shown as a black line in Fig. 3.9, is included as a white line here. We see that the laminar plot has a fractal structure, similar to Fig. 2.14 for the revtokamap, but we can clearly identify the fingers, which have very large connection lengthes compared to the areas in between. This does not mean that a field line passing through a dark colored area takes very long to reach the wall, because the rotation number is the sum of the forward and backward number of toroidal rotations a field line needs to hit the wall with both endings. A field line on the unstable manifold hits the wall very fast for forward iteration, but backwards the unstable manifold becomes the stable one and so the

Chapter 3. Cylindrical model for magnetic field lines in TEXTOR-DED

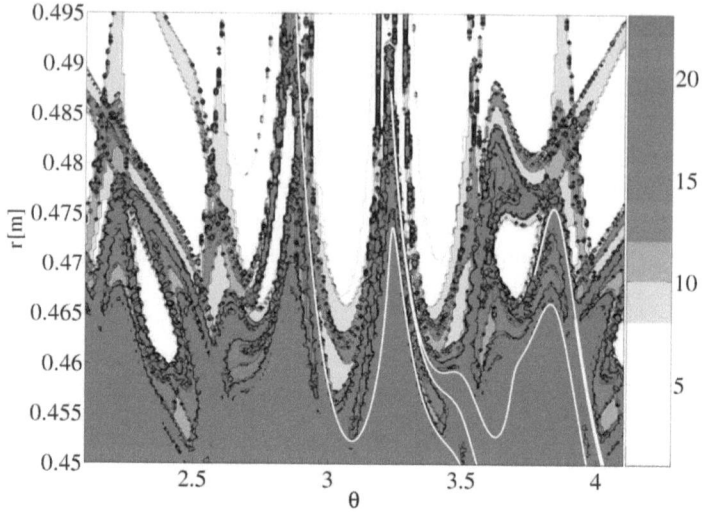

Figure 3.10: Laminar plot of the symmetric DED map with $I_0 = 10$ kA for the toroidal rotation number. Cutout of the area with the two main fingers around $\theta = \pi$. The white line is the black colored left sided unstable manifold of Fig. 3.9.

field line converges towards the hyperbolic point and will never hit the wall. This means that the field lines on the manifolds have infinite connection lengthes. Although the last resonance dominates the footprints, field lines from the finger areas are able to penetrate the plasma deeply and remain inside the plasma for a very long time. Through the fingers the hot inside of the plasma is connected to the wall. This shows that the heat deposition at the wall occurs at the finger positions. This will be further investigated on the real toroidal model in the next chapter, where we will also compare our results to measurements of the heat flux in TEXTOR.

Chapter 4

Toroidal DED model with relativistic particle drift effects

Up to now, we only considered the magnetic field lines of the DED in a cylindrical model. The next step is to modify the model for toroidal geometry [5]. We generalize our description of the DED to include also relativistic drift effects. The charged particles of the plasma are performing gyrations around the magnetic field lines. The gyro-radius scales with the magnetic field. Due to the main magnetic field of about $B_0 = 2$ T, this radius is very small compared to the dimensions of the torus. Because of the curvature and gradients of the magnetic field, the particles are also drifting, which is a much larger effect than the gyration. Therefore, we can neglect the gyration in comparison to the drift motion and consider the guiding-center only. The deviation of the drift surfaces from the magnetic KAM surfaces increases with the particle energy. So, the kinetic energy of the particles is, in addition to the perturbation current, one of the most important parameters of the particle system. The typical energy scale of the plasma particles ranges from a few keV up to 15 MeV and more. Especially the so called runaway electrons are extremely high energetic particles, which are of great interest. They cause losses of a lot of energy for the fusion system, due to their syncrotron radiation on the one hand and their possible escape from the system on the other hand. For such particles the kinetic energy is much larger than the rest energy, 0.511 MeV for electrons. So we have to describe the particle motion relativistically. A non-relativistic case has been discussed in [19, 20].

The model derived here includes magnetic field lines as well as particle drift effects, both in toroidal geometry. It describes particles with any charge, mass, kinetic energy and relative direction of motion (in or counter direction of the field lines). The theory also includes an arbitrary external electric field, but we will not evaluate electric field effects in detail. Note that the particles are still described as "test-particles", meaning that the electric and magnetic fields are not selfconsistent. Any influences of the particles on the fields are neglected. Further the model does not include trapped particles.

Chapter 4. Toroidal DED model with relativistic particle drift effects

Our first task is to construct a mapping procedure for the relativistic particle drift model in toroidal geometry. We start with the relativistic Hamiltonian of a particle in an external electromagnetic field.

4.1 The Hamiltonian for charged relativistic particles in an EM field

We assume a particle with the mass m_0 and the charge $q = Z_q e$, while Z_q is the charge number of the particle ($Z_q = -1$ for electrons) and e is the elementary charge, inside an electromagnetic (EM) field, described by the magnetic vector potential $\vec{A}$ and the electric scalar potential $\hat{\phi}$. The vector potential includes the magnetic equilibrium field as well as the perturbation field, while the electric field, given by $\vec{E} = -\nabla \hat{\phi}$ is only an external additional perturbation. We use cylindrical coordinates $(\hat{R}, \varphi, \hat{Z})$ to describe the toroidal geometry. In contrast to the cylindrical DED model the angle coordinate now represents the toroidal angle of the torus, while the radial coordinate corresponds to the major radius. Using these coordinates the Hamiltonian reads in relativistic form

$$\hat{H} = \left[m_0^2 c^4 + c^2 \left(\hat{p}_R - \frac{q}{c} A_R \right)^2 + \frac{c^2}{\hat{R}^2} \left(\hat{p}_\varphi - \frac{q}{c} \hat{R} A_\varphi \right)^2 + c^2 \left(\hat{p}_z - \frac{q}{c} A_z \right)^2 \right]^{1/2} + q\hat{\phi} . \quad (4.1.1)$$

Due to the gauge invariance of the vector potential, we can assume its radial coordinate A_R to be zero, i.e. $\vec{A} = (0, A_\varphi, A_z)$. The toroidal field is mainly determined by the A_z component, while the A_φ component describes the poloidal field and the perturbation field.

To describe the interior of the torus tube properly, we normalize the coordinates as follows

$$x = \frac{\hat{R} - R_0}{R_0} , \quad z = \frac{\hat{Z}}{R_0} , \quad t = \omega_c \hat{T} , \quad (4.1.2)$$

$$p_x = \frac{\hat{p}_R}{m_0 \omega_c R_0} , \quad p_\varphi = \frac{\hat{p}_\varphi}{m_0 \omega_c R_0^2} , \quad p_z = \frac{\hat{p}_z}{m_0 \omega_c R_0} , \quad (4.1.3)$$

using the major radius R_0 of the torus. The coordinates x and z represent a cartesian coordinate system perpendicular to the toroidal angle coordinate with its center at the geometrical center of the torus tube. To normalize the time scale we used the gyro-frequency

$$\omega_c = \frac{eB_0}{m_0 c} \quad (4.1.4)$$

with the main magnetic field B_0. Normalizing the Hamiltonian and the scalar potential

$$\tilde{H} = \frac{\hat{H}}{m_0 \omega_c^2 R_0^2} , \quad \phi = \frac{q}{m_0 \omega_c^2 R_0^2} \hat{\phi} , \quad (4.1.5)$$

62

4.1 The Hamiltonian for charged relativistic particles in an EM field

we obtain

$$\begin{aligned}\tilde{H} &= \frac{1}{m_0\omega_c^2 R_0^2}\Big[m_0^2 c^4 + c^2(m_0\omega_c R_0 p_x)^2 + \frac{c^2}{(R_0 x + R_0)^2}(m_0\omega_c R_0^2 p_\varphi - \frac{q}{c}(R_0 x + R_0)A_\varphi)^2 \\ &\quad + c^2(m_0\omega_c R_0 p_z - \frac{q}{c}A_z)^2\Big]^{1/2} + \phi \\ &= \left[\frac{c^4}{\omega_c^4 R_0^4} + \frac{c^2}{\omega_c^2 R_0^2}\left(p_x^2 + \left(\frac{p_\varphi}{1+x} - \frac{qA_\varphi}{cm_0\omega_c R_0}\right)^2 + \left(p_z - \frac{qA_z}{cm_0\omega_c R_0}\right)^2\right)\right]^{1/2} + \phi\,.\end{aligned}$$

Introducing the normalized particle energy at the rest

$$\varepsilon_0 = \frac{c^2}{\omega_c^2 R_0^2} \qquad (4.1.6)$$

and the normalized vector potential

$$f_\varphi = \frac{Z_q}{B_0 R_0}(1+x)A_\varphi, \quad f_z = \frac{Z_q}{B_0 R_0}A_z\,, \qquad (4.1.7)$$

the Hamiltonian reads

$$\tilde{H} = \left[\varepsilon_0^2 + \varepsilon_0\left(p_x^2 + \frac{(p_\varphi - f_\varphi)^2}{(1+x)^2} + (p_z - f_z)^2\right)\right]^{1/2} + \phi\,. \qquad (4.1.8)$$

It is much more convenient to expand the Hamiltonian to the 8-dimensional phase space $(q_i, p_i) = (x, \varphi, z, t, p_x, p_\varphi, p_z, p_t)$, $i = 1, \ldots, 4$, including the time t and the energy H as additional canonical variables, because for the mapping technique, we will have to change the independent variable to the toroidal angle φ. Defining the to the time t corresponding canonical momentum

$$p_t = -\tilde{H}\,, \qquad (4.1.9)$$

we can introduce

$$U = \frac{1}{2}\left(\frac{(p_\varphi - f_\varphi)^2}{(1+x)^2} + (p_z - f_z)^2 - \frac{(-p_t - \phi)^2}{\varepsilon_0} + \varepsilon_0\right) \qquad (4.1.10)$$

as an effective potential for a one dimensional particle motion, described by the Hamiltonian

$$H = \frac{1}{2}\left(p_x^2 + \frac{(p_\varphi - f_\varphi)^2}{(1+x)^2} + (p_z - f_z)^2 - \frac{(-p_t - \phi)^2}{\varepsilon_0} + \varepsilon_0\right) = \frac{1}{2}p_x^2 + U = 0\,. \qquad (4.1.11)$$

Such a formulation will be needed to transform to guiding center coordinates, which are used to neglect the fast and small gyration of the particles around the field lines. This will be done in the next section. Up to now, the dynamics of the whole Hamiltonian system in the extended

8-dimensional phase space, including the gyration, is given by the canonical equations of motion

$$\frac{dq_i}{d\tau} = \frac{\partial H}{\partial p_i}, \quad \frac{dp_i}{d\tau} = -\frac{\partial H}{\partial q_i} \tag{4.1.12}$$

with the time τ, which is the relativistic reference time within the co-moving frame of reference.

4.2 Guiding-center approximation

The Hamiltonian (4.1.11) describes the complete dynamics of a particle within the electromagnetic field, including the fast gyration of the particles around the field lines. The direct numerical integration of the equations of motion would require very long computational time, because one would have to choose step sizes small enough to resolve the particle gyration. To simplify the problem significantly, we transform the fast varying variables $(x, \varphi, z, t, p_x, p_\varphi, p_z, p_t)$ to the slow varying guiding center coordinates $(\vartheta_x, \Phi, Z, T, I_x, P_\varphi, P_z, P_t)$ to eliminate the fast gyration of the particles. The transformation is generated by the function

$$F = \varphi P_\varphi + z P_z + t P_t + \varepsilon S(x, I_x, \varphi, P_\varphi, z, P_z, t, P_t), \tag{4.2.1}$$

which depends on the old coordinates and the new momentums, with

$$\varepsilon S = \int p_x(x', I_x, \varphi, P_\varphi, z, P_z, t, P_t)\, dx', \tag{4.2.2}$$

whereas ε is a small parameter, given by the relation ρ_x/L of the gyro-radius ρ_x to the characteristic scale L of the system. We use (4.2.2) to transform the radial coordinates (x, p_x) to the action-angle variables (ϑ_x, I_x), given by

$$I_x = \frac{1}{2\pi} \oint_C p_x(x)\, dx, \quad \vartheta_x = \varepsilon \frac{\partial S}{\partial I_x}. \tag{4.2.3}$$

The integration is taken along the contour C of one full-turn gyration of the particle in the (x, p_x) plane.

The guiding-center is determined by the minimum of the effective potential $U(x)$, Eq. (4.1.10). The radial dependency of U is plotted for various electron energies in Fig. 4.1. The expansion of U clearly shows the guiding center position and the width of the gyro oscillations. This figure is similar to figure 1 of [19], where the effective potential for the non-relativistic case is shown. Note that the potential is shifted by the total energy $E = \gamma \varepsilon_0$ to make it more comparable with the non-relativistic case, where H is the total energy. Here we have $H = 0$, because we have subtracted the total energy.

4.2 Guiding-center approximation

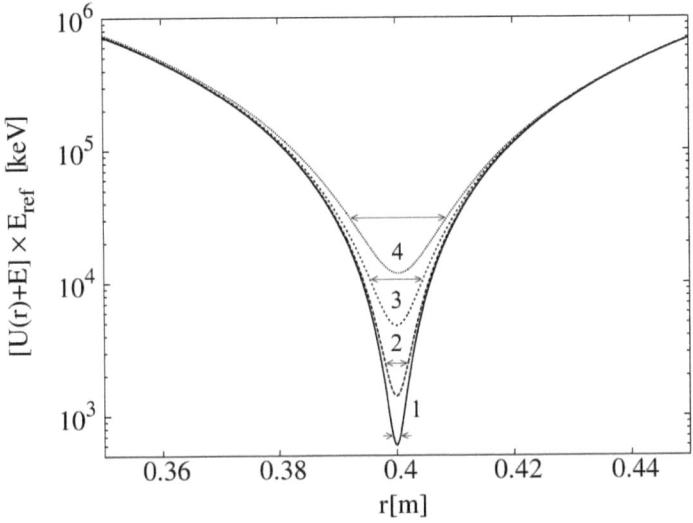

Figure 4.1: Radial dependency of the effective potential $U(r)$, Eq. (4.1.10), shifted by the total Energy $E = \gamma \varepsilon_0$, for various kinetic electron energies: 1. 0.1 MeV, 2. 1 MeV, 3. 5 MeV and 4. 15 MeV. The position and width of the gyro oscillation is marked by the arrows. The tokamak field is described by the equilibrium field. E_{ref} is the reference energy, used for normalization, $r = R_0 x$, $R_0 = 1.75$ m.

Due to the smallness of the gyro-radius, marked by the arrows in Fig. 4.1, we can expand the potential in a Taylor series around the guiding-center position x_c and get

$$H = \frac{1}{2}p_x^2 + U(x_c) + \frac{1}{2}\omega_x^2(x - x_c)^2 + O((x - x_c)^3) \,. \tag{4.2.4}$$

The minimum is given by the zero point of the first derivative with respect to x of the potential, while the second derivative is lager than zero. We demand

$$\left. \frac{\partial U}{\partial x} \right|_{x=x_c} = 0 \tag{4.2.5}$$

$$\Rightarrow \frac{(p_\varphi - f_\varphi)^2}{(1+x)^3} + \frac{p_\varphi - f_\varphi}{(1+x)^2}\frac{\partial f_\varphi}{\partial x} + (p_z - f_z)\frac{\partial f_z}{\partial x} - \frac{-p_t - \phi}{\varepsilon_0}\frac{\partial \phi}{\partial x} = 0 \tag{4.2.6}$$

$$\Leftrightarrow u_\varphi^2 + u_\varphi \frac{\partial f_\varphi}{\partial x} + (1+x)u_z\frac{\partial f_z}{\partial x} - (1+x)\gamma\frac{\partial \phi}{\partial x} = 0 \tag{4.2.7}$$

Chapter 4. Toroidal DED model with relativistic particle drift effects

with
$$u_\varphi = \frac{p_\varphi - f_\varphi}{1+x}, \quad u_z = p_z - f_z, \quad \gamma = \frac{-p_t - \phi}{\varepsilon_0}. \tag{4.2.8}$$

The solution $x_c = x_c(\varphi, z, t, p_\varphi, p_z, p_t)$ of Eq. (4.2.7) is the position of the guiding-center. The second derivative at x_c defines the frequency of the particle, which oscillates around the guiding-center, the gyro-frequency

$$\omega_x(x_c, \Phi, Z, T, I_x, P_\varphi, P_z, P_t) = \left. \frac{\partial^2 U}{\partial x^2} \right|_{x=x_c}. \tag{4.2.9}$$

After determining the guiding-center position and the gyro-frequency, we can perform the transformation (4.2.1) with Eq. (4.2.2). Therefore, we expand the Hamiltonian into a power series of ε and obtain for the Hamiltonian in its transformed form

$$\begin{aligned}
H &= H_0 + \varepsilon H_1(\vartheta_x, \Phi, Z, T, I_x, P_\varphi, P_z, P_t) + O(\varepsilon^2) \\
&= \omega_x(x_c, \Phi, Z, T, I_x, P_\varphi, P_z, P_t) I_x + U(x_c, \Phi, Z, T, I_x, P_\varphi, P_z, P_t) \\
&\quad + \varepsilon H_1(\vartheta_x, \Phi, Z, T, I_x, P_\varphi, P_z, P_t) + O(\varepsilon^2) = 0.
\end{aligned} \tag{4.2.10}$$

The zeroth order in ε then describes the guiding-center motion, which does not depend on the fast gyro-phase ϑ_x. As shown in [19], the action variable I_x is a good adiabatic invariant of motion for typical tokamak plasmas. Also the gyro-frequency ω_x is much larger than the toroidal and poloidal transit frequencies of the drift motion. Therefore, we can neglect the first and higher orders of ε, which are containing the gyro-phase ϑ_x.

To find the relations between the geometrical and guiding-center coordinates, we have to determine the unknown part S of the generating function. For this we resolve Eq. (4.2.4) with respect to p_x

$$p_x = \sqrt{2(H - U(x_c)) - \omega_x^2 (x - x_c)^2}, \tag{4.2.11}$$

and get from Eq. (4.2.2)

$$\varepsilon S = \int \sqrt{2 I_x \omega_x - \omega_x^2 (x' - x_c)^2} \, dx' = \int \sqrt{2 I_x \omega_x} \sqrt{1 - \frac{\omega_x}{2 I_x}(x' - x_c)^2} \, dx'. \tag{4.2.12}$$

Introducing the substitution

$$y = \frac{x' - x_c}{\sqrt{2 I_x / \omega_x}} \quad \Rightarrow \quad dx' = \sqrt{\frac{2 I_x}{\omega_x}} dy, \tag{4.2.13}$$

we find

$$\varepsilon S = 2 I_x \int \sqrt{1 - y^2} \, dy = I_x \left[\arcsin y + y \sqrt{1 - y^2} \right]. \tag{4.2.14}$$

4.3 Simplification of the guiding-center equations

The to the action-variable I_x corresponding angle variable ϑ_x reads

$$\vartheta_x = \frac{\partial S}{\partial I_x} = \arcsin y + y\sqrt{1-y^2} + 2I_x\sqrt{1-y^2}\frac{x-x_c}{\sqrt{2/\omega_x}}I_x^{-3/2}(-\frac{1}{2}) = \arcsin y \,. \quad (4.2.15)$$

The relations between the old coordinates (q_i, p_i) and the guiding-center coordinates (Q_i, P_i) are then given by the derivatives of the generating function (4.2.1)

$$Q_i = \frac{\partial F}{\partial P_i}\,, \quad p_i = \frac{\partial F}{\partial q_i}\,, \quad i = 1,\ldots,4\,, \quad (4.2.16)$$

thus

$$x = x_c + \sqrt{\frac{2I_x}{\omega_x}}\sin\vartheta_x\,, \quad p_x = \sqrt{2I_x\omega_x}\cos\vartheta_x\,, \quad (4.2.17)$$

$$q_i = Q_i + \sqrt{2I_x\omega_x}\frac{\partial x_c}{\partial P_i}\cos\vartheta_x - \frac{I_x}{\omega_x}\frac{\partial \omega_x}{\partial P_i}\sin 2\vartheta_x\,, \quad (4.2.18)$$

$$p_i = P_i - \sqrt{2I_x\omega_x}\frac{\partial x_c}{\partial q_i}\cos\vartheta_x + \frac{I_x}{\omega_x}\frac{\partial \omega_x}{\partial q_i}\sin 2\vartheta_x\,, \quad (4.2.19)$$

whereas the action variable I_x is considered as a constant of motion.

4.3 Simplification of the guiding-center equations

We can simplify the equations for the guiding-center (4.2.7), (4.2.9) and the relations between the real and guiding-center coordinates significantly by neglecting all perturbations to the toroidal magnetic main field $B_\varphi = B_0 R_0/\hat{R}$. This leads to the ansatz

$$A_z(\hat{R}) = -B_0 R_0 \ln\frac{\hat{R}}{R_0} \quad \Rightarrow \quad f_z = -Z_q \ln(1+x) \quad (4.3.1)$$

for the z component of the vector potential. This ansatz is valid, because B_0 is much larger than any perturbation of the DED coils. We also can neglect the terms in Eq. (4.2.7) which are proportional to u_φ^2, $u_\varphi \frac{\partial f_\varphi}{\partial x}$ and $\frac{\partial \phi}{\partial x}$. For relativistic electrons we have $u \approx 10^{-2}$, while all other field components are small compared to B_0, according to [19]. Then we obtain from Eq. (4.2.7)

$$(1+x)(p_z - f_z)\frac{\partial f_z}{\partial x} = 0 \quad \Rightarrow \quad p_z = f_z \quad \Rightarrow \quad x_c = e^{-p_z/Z_q} - 1\,. \quad (4.3.2)$$

With this result, we can determine the gyro-frequency, according to Eq. (4.2.9)

$$\omega_x = \frac{1}{1+x_c} = e^{p_z/Z_q}\,. \quad (4.3.3)$$

Chapter 4. Toroidal DED model with relativistic particle drift effects

The dependencies of the real coordinates and the guiding-center coordinates simplify to

$$x = x_c + \sqrt{\frac{2I_x}{\omega_x}} \sin \vartheta_x , \quad p_x = \sqrt{2I_x\omega_x} \cos \vartheta_x , \quad (4.3.4)$$

$$z = Z + \sqrt{\frac{2I_x}{\omega_x}} \cos \vartheta_x , \quad p_z = P_z , \quad (4.3.5)$$

$$\varphi = \Phi , \quad p_\varphi = P_\varphi , \quad t = T , \quad p_t = P_t . \quad (4.3.6)$$

Neglecting the fast gyro-phase, the Hamiltonian (4.2.10) reads

$$H = \omega_x(p_z)I_x + \frac{1}{2}\left[\frac{(p_\varphi - f_\varphi(x_c, \varphi, z, t))^2}{(1 + x_c)^2} - \frac{(-p_t - \phi(x_c, \varphi, z, t))^2}{\varepsilon_0} + \varepsilon_0\right] = 0 , \quad (4.3.7)$$

while the particle dynamics are given by the Hamiltonian equations of motion

$$\dot{z} = I_x \frac{\partial \omega_x}{\partial p_z} + \frac{\partial x_c}{\partial p_z}\left[-\frac{(p_\varphi - f_\varphi)}{(1 + x_c)^2}\frac{\partial f_\varphi}{\partial x_c} - \frac{(p_\varphi - f_\varphi)^2}{(1 + x_c)^3} + \frac{-p_t - \phi}{\varepsilon_0}\frac{\partial \phi}{\partial x_c}\right] \quad (4.3.8)$$

$$\dot{p}_z = \frac{(p_\varphi - f_\varphi)}{(1 + x_c)^2}\frac{\partial f_\varphi}{\partial z} - \frac{-p_t - \phi}{\varepsilon_0}\frac{\partial \phi}{\partial z} \quad (4.3.9)$$

$$\dot{\varphi} = \frac{(p_\varphi - f_\varphi)}{(1 + x_c)^2} \quad \dot{p}_\varphi = \frac{(p_\varphi - f_\varphi)}{(1 + x_c)^2}\frac{\partial f_\varphi}{\partial \varphi} - \frac{-p_t - \phi}{\varepsilon_0}\frac{\partial \phi}{\partial \varphi} \quad (4.3.10)$$

$$\dot{t} = \frac{-p_t - \phi}{\varepsilon_0} = \gamma \quad \dot{p}_t = \frac{(p_\varphi - f_\varphi)}{(1 + x_c)^2}\frac{\partial f_\varphi}{\partial t} - \frac{-p_t - \phi}{\varepsilon_0}\frac{\partial \phi}{\partial t} \quad (4.3.11)$$

$$\dot{\vartheta}_x = \omega_x(p_z) \quad \dot{I}_x = 0 \quad (4.3.12)$$

with the derivative $\dot{q} = \frac{dq}{d\tau}$ with respect to the reference time τ.

For the mapping it is much more convenient to reformulate the Hamiltonian description. We introduce the toroidal angle φ as the independent time-like variable and the new Hamiltonian

$$K = -p_\varphi , \quad (4.3.13)$$

given by the corresponding canonical momentum. From Eq. (4.3.7) we find

$$\omega_x I_x + \frac{1}{2}\left[\varepsilon_0 - \frac{(-p_t - \phi)^2}{\varepsilon_0}\right] = -\frac{1}{2}\frac{(-K - f_\varphi)^2}{(1 + x_c)^2} \quad (4.3.14)$$

$$\Leftrightarrow -2(1 + x_c)^2\omega_x I_x - (1 + x_c)^2\left[\varepsilon_0 - \frac{(-p_t - \phi)^2}{\varepsilon_0}\right] = (-K - f_\varphi)^2 \quad (4.3.15)$$

$$\Leftrightarrow \pm(1 + x_c)\sqrt{-2\omega_x I_x - \varepsilon_0 + \frac{(-p_t - \phi)^2}{\varepsilon_0}} = -K - f_\varphi \quad (4.3.16)$$

$$\Rightarrow K = -f_\varphi - \sigma(1 + x_c)\left[\varepsilon_0(\gamma^2 - 1) - 2\omega_x I_x\right]^{1/2} . \quad (4.3.17)$$

Here we introduced the new parameter $\sigma = \pm 1$, which determines the direction of motion relatively to the field lines. For $\sigma = 1$ the particles are moving in the direction of the field lines and we are talking about co-passing particles. For $\sigma = -1$ the particles are moving in the opposite direction, the so called counter-passing particles. The dynamics are given by the Hamiltonian equations of motion, now with respect to the new Hamiltonian K,

$$\frac{dz}{d\varphi} = \frac{1}{Z_q}(1+x_c)\left[\frac{\partial f_\varphi}{\partial x_c} + \sigma\left(\varepsilon_0(\gamma^2-1) - \omega_x I_x - (1+x_c)\gamma\frac{\partial \phi}{\partial x_c}\right)\right.$$
$$\left. \times \left(\varepsilon_0(\gamma^2-1) - 2\omega_x I_x\right)^{-1/2}\right] \quad (4.3.18)$$

$$\frac{dp_z}{d\varphi} = \frac{\partial f_\varphi}{\partial z} - \sigma(1+x_c)\gamma\frac{\partial \phi}{\partial z}\left(\varepsilon_0(\gamma^2-1) - 2\omega_x I_x\right)^{-1/2} \quad (4.3.19)$$

$$\frac{dt}{d\varphi} = \sigma(1+x_c)\gamma\left(\varepsilon_0(\gamma^2-1) - 2\omega_x I_x\right)^{-1/2} \quad (4.3.20)$$

$$\frac{dp_t}{d\varphi} = \frac{\partial f_\varphi}{\partial t} - \sigma(1+x_c)\gamma\frac{\partial \phi}{\partial t}\left(\varepsilon_0(\gamma^2-1) - 2\omega_x I_x\right)^{-1/2} \quad (4.3.21)$$

$$\frac{d\vartheta_x}{d\varphi} = \sigma\left(\varepsilon_0(\gamma^2-1) - 2\omega_x I_x\right)^{-1/2} \quad (4.3.22)$$

$$\frac{dI_x}{d\varphi} = 0. \quad (4.3.23)$$

4.4 The equilibrium field

The z-component of the vector potential, defining the toroidal magnetic field has already been introduced in the previous section. Now we model the rest of the equilibrium field, given by the unperturbed toroidal component $A_\varphi^{(0)}$ of the vector potential

$$A_\varphi^{(0)} = \frac{B_0}{\hat{R}} \int \frac{d\psi}{q(\rho(\psi))} \quad \Rightarrow \quad f_\varphi^{(0)} = \frac{Z_q}{R_0^2} \int \frac{d\psi}{q(\rho(\psi))}, \quad (4.4.1)$$

whereas $q(\rho)$ is the safety factor. The safety factor is specified in Sec. 3.3 as profile "b", given by Eq. (3.3.11) and the dashed line of Fig. 3.3. $\psi = \frac{1}{2}\rho^2$ is the toroidal magnetic flux, dealing as action variable and ρ is the minor radius of the magnetic KAM surfaces. Due to the toroidal geometry and the plasma, the magnetic surfaces are shifted, compared to the spheres with the geometrical radius $r^2 = \sqrt{x^2 + z^2}$. This shift is the so called Shafranov shift [33]

$$\Delta(\rho) = R(\rho) - R_a \approx (\Lambda + 1)\frac{a^2 - \rho^2}{2R_a}, \quad (4.4.2)$$

with the minor radius of the plasma a, the major radius of the plasma center $R_a = R(a)$ and the parameter $\Lambda = \beta_{pol} + l_i/2 - 1$. This parameter includes β_{pol}, the ratio of the plasma pressure to the pressure of the poloidal magnetic field, and the internal inductance l_i. $R(\rho)$ is the major

Chapter 4. Toroidal DED model with relativistic particle drift effects

radius of the center of the surface with the minor radius ρ. With the Shafranov shift the relation between the minor radius ρ and the geometrical coordinates reads

$$\rho = \sqrt{(\hat{R} - R_a - \Delta(\rho))^2 + \hat{Z}^2} \ . \tag{4.4.3}$$

Because the Shafranov shift itself depends on the shifted radius ρ, Eq. (4.4.3) is an implicit equation, which can be resolved with respect to ρ, as shown in Sec. 4.5.

First, we normalize all length-like variables and parameters, using the major radius R_0 of the torus. For simplicity, we keep the existing notations and get

$$\frac{R_a}{R_0} \to R_a \ , \quad \frac{a}{R_0} \to a \ , \quad \frac{\rho}{R_0} \to \rho \ , \quad \frac{\psi}{R_0^2} = \frac{\rho^2}{2R_0^2} \to \psi = \frac{1}{2}\rho^2 \ . \tag{4.4.4}$$

The normalizations of $\hat{R}$ and $\hat{Z}$ are already given by Eq. (4.1.2). We further obtain

$$\frac{\Delta(\rho)}{R_0} = (\Lambda + 1)\frac{a^2/R_0^2 - \rho^2/R_0^2}{2R_a/R_0} \to \Delta(\rho) = (\Lambda + 1)\frac{a^2 - \rho^2}{2R_a} \tag{4.4.5}$$

and therefore

$$\rho \to \rho = \sqrt{(1 + x - R_a - \Delta(\rho))^2 + z^2} \ , \tag{4.4.6}$$

$$f_\varphi^{(0)} \to f_\varphi^{(0)} = Z_q \int \frac{d\psi}{q(\rho(\psi))} \ . \tag{4.4.7}$$

For the Hamiltonian equations of motion (4.3.18)-(4.3.23) we need the derivatives

$$\frac{\partial f_\varphi^{(0)}}{\partial q_i} = Z_q \frac{\rho}{q(\rho)} \frac{\partial \rho}{\partial q_i} \tag{4.4.8}$$

with respect to the coordinates $q_i = (x, z, t)$. For this we resolve Eq. (4.4.6) with respect to the coordinates q_i

$$x = R_a + \Delta + \sqrt{\rho^2 - z^2} - 1 \ \Rightarrow \ \frac{\partial x}{\partial \rho} = \frac{\partial \Delta}{\partial \rho} + \frac{\rho}{1 + x - R_a - \Delta} \ , \tag{4.4.9}$$

$$z = \sqrt{\rho^2 - (1 + x - R_a - \Delta)^2} \ \Rightarrow \ \frac{\partial z}{\partial \rho} = \frac{1}{z}(\rho + (1 + x - R_a - \Delta)\frac{\partial \Delta}{\partial \rho}) \ . \tag{4.4.10}$$

Insert this result in Eq. (4.4.8), we find

$$\frac{\partial f_\varphi^{(0)}}{\partial x} = \frac{Z_q}{q(\rho)} \frac{1+x-R_a-\Delta}{(1+x-R_a-\Delta)\frac{1}{\rho}\frac{\partial\Delta}{\partial\rho}+1}, \qquad (4.4.11)$$

$$\frac{\partial f_\varphi^{(0)}}{\partial z} = \frac{Z_q}{q(\rho)} \frac{z}{(1+x-R_a-\Delta)\frac{1}{\rho}\frac{\partial\Delta}{\partial\rho}+1}, \qquad (4.4.12)$$

$$\frac{\partial f_\varphi^{(0)}}{\partial t} = 0. \qquad (4.4.13)$$

4.5 Explicit solution for ρ

For the numerics it is essential to determine the shifted minor radius ρ from the geometrical coordinates. The relation between ρ, x and z is given by Eq. (4.4.6), but this equation is an implicit one, because the Shafranov shift Δ also depends on ρ. Now we derive an explicit equation for ρ. Therefore, we consider

$$\rho^2 = (1+x-R_a-\Delta)^2 + z^2 \quad \text{with} \quad \Delta = \frac{\Lambda+1}{2R_a}(a^2-\rho^2). \qquad (4.5.1)$$

Introducing the shortcut

$$\rho_0^2 = (1+x-R_a)^2 + z^2, \qquad (4.5.2)$$

we obtain

$$\begin{aligned}
0 &= \frac{(\Lambda+1)^2}{4R_a^2}(a^4+\rho^4-2a^2\rho^2) - (1+x-R_a)\frac{\Lambda+1}{R_a}(a^2-\rho^2) - \rho^2 + \rho_0^2 \\
&= \rho^4 + \frac{4R_a^2}{(\Lambda+1)^2}\left[-(1+x-R_a)\frac{\Lambda+1}{R_a}(a^2-\rho^2) - \rho^2 + \rho_0^2\right] - 2a^2\rho^2 + a^4 \\
&= \rho^4 + \left\{\frac{4R_a^2}{(\Lambda+1)^2}\left[(1+x-R_a)\frac{\Lambda+1}{R_a}-1\right] - 2a^2\right\}\rho^2 \\
&\quad + a^4 + \frac{4R_a^2}{(\Lambda+1)^2}\left[-(1+x-R_a)\frac{\Lambda+1}{R_a}a^2 + \rho_0^2\right].
\end{aligned}$$

Defining

$$2p(x,z) = \frac{4R_a^2}{(\Lambda+1)^2}\left[(1+x-R_a)\frac{\Lambda+1}{R_a}-1\right] - 2a^2, \qquad (4.5.3)$$

$$q(x,z) = a^4 + \frac{4R_a^2}{(\Lambda+1)^2}\left[-(1+x-R_a)\frac{\Lambda+1}{R_a}a^2 + \rho_0^2\right], \qquad (4.5.4)$$

the explicit solutions for ρ are given by

$$\rho_{1/2} = \sqrt{-p \pm \sqrt{p^2 - q}} \geq 0. \qquad (4.5.5)$$

As one can see, there are two possible solutions. Assuming that both solutions really exist, which would only be possible, if $\rho^2 \geq 0$ and real for $+\sqrt{p^2 - q}$ and $-\sqrt{p^2 - q}$, the solution closest to ρ_0 would be the proper choice. But typically there is only one real solution.

4.6 The perturbation field

The last component we have to include is the perturbation field of the DED. According to Sec. 3.2 and [5] the vector potential of the DED field reads

$$A_\varphi^{(1)} = (1+x)^{-1/2} \sum_{m=0}^{\infty} G_m \left(\frac{r}{r_c}\right)^m \cos(m\theta - n_0\varphi + \omega t) \qquad (4.6.1)$$

with the Fourier modes

$$G_m = (-1)^{m+1} \mu_0 I_0 \frac{m_0}{m\pi} g_m \quad \text{with} \quad g_m = \frac{\sin((m-m_0)n_0\pi/m_0)}{(m-m_0)\pi}, \qquad (4.6.2)$$

while $r = \sqrt{x^2 + z^2}$ and $\theta = \arctan(z/x)$. Note that we are using the cosine here instead of the sine, used in Sec. 3.2, for more consistency with the work on the non-relativistic case [19]. The physics are the same, because we only take another toroidal position for the Poincaré section as for the DED map.

Performing the normalization (4.1.7), we get

$$f_\varphi^{(1)} = Z_q \frac{1+x}{B_0 R_0} A_\varphi^{(1)}. \qquad (4.6.3)$$

By superposition with the equilibrium field, the total normalized toroidal vector potential reads

$$f_\varphi = f_\varphi^{(0)} + f_\varphi^{(1)}. \qquad (4.6.4)$$

4.7 The mapping procedure for the relativistic drift model

Considering the results of the previous sections, we have all necessary parts to construct the mapping procedure for the relativistic drift model. We write the Hamiltonian (4.3.17) in the following form

$$K = K_0(z, t, p_z, p_t) + K_1(z, t, p_z, p_t, \varphi) \qquad (4.7.1)$$

4.7 The mapping procedure for the relativistic drift model

with

$$K_0(z,t,p_z,p_t) = -f_\varphi^{(0)}(x_c,z) - \sigma(1+x_c)\left[\varepsilon_0(\gamma^2-1) - 2\omega_x I_x\right]^{1/2}, \qquad (4.7.2)$$

describing the unperturbed relativistic motion of the particle guiding center, and

$$K_1(\varphi,z,t,p_z,p_t) = -f_\varphi^{(1)}(x_c,\varphi,z,t), \qquad (4.7.3)$$

describing the DED perturbation, which is small compared to K_0. In the absence of the magnetic perturbation and an electric field, $\phi = 0$, there are two independent constants of motion. We introduce the action variable

$$I_z = \frac{1}{2\pi} \int_{C_z} p_z(z)\, dz, \qquad (4.7.4)$$

which is a constant of motion. The integration is taken along the projection of the unperturbed drift orbit onto the (x,p_z) plane. The second invariant is the particle energy $I_t = -H = p_t$. We further introduce the corresponding angle variables

$$\vartheta_z = \frac{\partial F}{\partial I_z} \quad \text{and} \quad \vartheta_t = \frac{\partial F}{\partial I_t} \qquad (4.7.5)$$

where $F = tI_t + \int p_z(z, I_z, I_t)\, dz$ is the corresponding generating function for the canonical transformation on these action-angle variables. In the new variables the Hamiltonian (4.7.1) reads

$$K = K_0(I_z, I_t) + K_1(I_z, \vartheta_z, I_t, \vartheta_t, \varphi) \qquad (4.7.6)$$

Now we expand the perturbation part K_1 into a Fourier series with respect to ϑ_z, according to Eq. (4.6.1),

$$K_1 = \sum_m K_m(I_z, I_t) \cos(m\vartheta_z - n_0\varphi + \omega\vartheta_t), \qquad (4.7.7)$$

with the Fourier coefficients

$$K_m = \frac{1}{2\pi} \int_0^{2\pi} K_1(I_z, \vartheta_z, I_t, \vartheta_t, \varphi) \cos(m\vartheta_z)\, d\vartheta_z. \qquad (4.7.8)$$

Then we can derive the generating function S for the mapping, according to Sec. 2.2, by integrating the perturbation part K_1 over φ along the unperturbed trajectories

$$\vartheta_z = \vartheta_{z0} + \Omega_z(I_z, I_t)(\varphi - \varphi_0), \qquad (4.7.9)$$
$$\vartheta_t = \vartheta_{t0} + \Omega_t(I_z, I_t)(\varphi - \varphi_0), \qquad (4.7.10)$$

Chapter 4. Toroidal DED model with relativistic particle drift effects

with the transit frequencies

$$\Omega_z(I_z, I_t) = \frac{\partial K_0}{\partial I_z} \quad \text{and} \quad \Omega_t(I_z, I_t) = \frac{\partial K_0}{\partial I_t} \ . \tag{4.7.11}$$

We obtain

$$S = -\int_{\varphi_0}^{\varphi} K_1(I_z, \vartheta_z(\varphi'), I_t, \vartheta_t(\varphi'), \varphi') \, d\varphi' \tag{4.7.12}$$

$$= -\sum_m K_m(I_z, I_t) \left[h_1(\eta) \sin \alpha + h_2(\eta) \cos \alpha \right] \tag{4.7.13}$$

with

$$\eta = m\Omega_z - n_0 + \omega\Omega_t \ , \quad \alpha = m\vartheta_z - n_0\varphi + \omega\vartheta_t \ , \quad k = \varphi - \varphi_0 \tag{4.7.14}$$

and

$$h_1(\eta) = \frac{1 - \cos(k\eta)}{\eta} \to 0 \quad \text{for } \eta \to 0 \ , \quad h_2(\eta) := \frac{\sin(k\eta)}{\eta} \to k \quad \text{for } \eta \to 0 \ . \tag{4.7.15}$$

Using this generating function, we can perform a 4-dimensional mapping

$$J_{z,k} = I_{z,k} - \frac{\partial S_k}{\partial \vartheta_{z,k}} \ , \qquad \bar{\vartheta}_{z,k} = \vartheta_{z,k} + \frac{\partial S_k}{\partial J_{z,k}} \ , \tag{4.7.16}$$

$$J_{t,k} = I_{t,k} - \frac{\partial S_k}{\partial \vartheta_{t,k}} \ , \qquad \bar{\vartheta}_{t,k} = \vartheta_{t,k} + \frac{\partial S_k}{\partial J_{t,k}} \ , \tag{4.7.17}$$

$$\bar{\vartheta}_{z,k+1} = \bar{\vartheta}_{z,k} + \Omega_z(J_{z,k}, J_{t,k})(\varphi_{k+1} - \varphi_k) \tag{4.7.18}$$

$$\bar{\vartheta}_{t,k+1} = \bar{\vartheta}_{t,k} + \Omega_t(J_{z,k}, J_{t,k})(\varphi_{k+1} - \varphi_k) \tag{4.7.19}$$

$$I_{z,k+1} = J_{z,k} + \frac{\partial S_{k+1}}{\partial \vartheta_{z,k+1}} \ , \qquad \vartheta_{z,k+1} = \bar{\vartheta}_{z,k+1} - \frac{\partial S_{k+1}}{\partial J_{z,k}} \ , \tag{4.7.20}$$

$$I_{t,k+1} = J_{t,k} + \frac{\partial S_{k+1}}{\partial \vartheta_{t,k+1}} \ , \qquad \vartheta_{t,k+1} = \bar{\vartheta}_{t,k+1} - \frac{\partial S_{k+1}}{\partial J_{t,k}} \ , \tag{4.7.21}$$

with $S_k = S(J_{z,k}, \vartheta_{z,k}, J_{t,k}, \vartheta_{t,k}, \varphi_k)$ and $S_{k+1} = S(J_{z,k}, \vartheta_{z,k+1}, J_{t,k}, \vartheta_{t,k+1}, \varphi_{k+1})$. We are using the symmetric mapping form with $\varphi_0 = \frac{1}{2}(\varphi_k + \varphi_{k+1})$.

In contrast to the mappings we have handled before, this mapping is 4-dimensional, because of the explicit time dependence. If we assume that the perturbation is static, meaning $\omega = 0$, the mapping reduces to the well known 2-dimensional one, because then the particle energy I_t is a constant of motion. In the following we concentrate on the 2-dimensional case with constant particle energy. This mapping reads

$$\xi_k = \psi_k - \varepsilon\frac{\partial S_k}{\partial \theta_k}, \qquad \vartheta_k = \theta_k + \varepsilon\frac{\partial S_k}{\partial \xi_k} \qquad (4.7.22)$$

$$\vartheta_{k+1} = \vartheta_k + \Omega(\xi_k)(\varphi_{k+1} - \varphi_k) \qquad (4.7.23)$$

$$\psi_{k+1} = \xi_k + \varepsilon\frac{\partial S_{k+1}}{\partial \theta_{k+1}}, \qquad \theta_{k+1} = \vartheta_{k+1} - \varepsilon\frac{\partial S_{k+1}}{\partial \xi_k} \qquad (4.7.24)$$

with $\psi = I_z$, $\xi = J_z$, $\theta = \vartheta_z$, $\vartheta = \bar{\vartheta}_z$ and $\Omega = \Omega_z$. The S_k and S_{k+1} are the same as above.

The main problem is now to determine the Fourier modes K_m and the transit frequencies Ω_z and Ω_t. Due to the fact that analytical calculations are highly non-trivial, we determine them only numerically by integrating the Hamiltonian equations of motion (4.3.18)-(4.3.23) for one full poloidal turn on the (x,z) plane. The numerics provide us with the results for the Fourier modes, the transit frequencies and their derivatives on a grid. Values between two grid points are interpolated, using cubic splines. The Runge-Kutta integrator and the cubic spline routines are taken from the numerical recipes [36]. A numerical code for the 4-dimensional mapping procedure has also been developed in the framework of this thesis, but we will not evaluate explicit time dependent perturbations here. The static case has to be analyzed in detail first.

4.8 Unperturbed drift surfaces with varying kinetic energy

As a first step, we concentrate on the unperturbed case to analyze, how the surfaces of constant q for particles, the so called drift surfaces, are shifted due to drift effects, compared to the magnetic KAM surfaces. Using the same q-profile, the shape of the drift surfaces depends on three parameters: the type of particles, the direction of the velocities and the kinetic energy of the particles.

Figure 4.2 shows various drift surfaces for different kinetic energies in comparison to the magnetic KAM surface for the field lines. All of them are surfaces with the same constant irrational value $q = \pi$. From the cylindrical model of Sec. 3 one could have expected that for the unperturbed magnetic surfaces r would be constant, but here the magnetic surface, given by the black line, is shifted outwards, compared to the geometrical center of the torus tube. This is caused by the Shafranov shift, which takes the effects of the toroidal geometry into account. Note that the poloidal angle $\theta = \pi$ marks the inner side of the torus, where the DED coils are mounted, while $\theta = 0$ corresponds to the outer side. As one can see from Fig. 4.2, the drift surfaces for co-passing electrons are also shifted outwards. The shift is larger for higher energies and always larger than the shift of the magnetic surface, which could be interpreted as the drift surface of particles with zero energy. This can be seen more clearly in Fig. 4.4, where the surfaces of Fig. 4.2 are shown, but now in a polar plot. The dashed line at the edge of the polar plot marks the torus wall at $r = 0.477$ m. The coils are located at the left side of the

Chapter 4. Toroidal DED model with relativistic particle drift effects

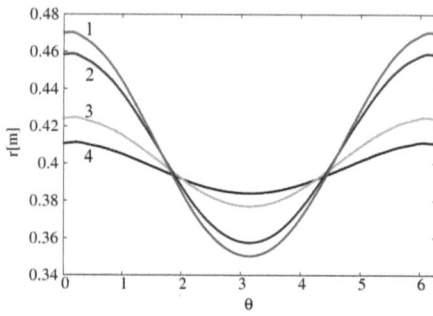 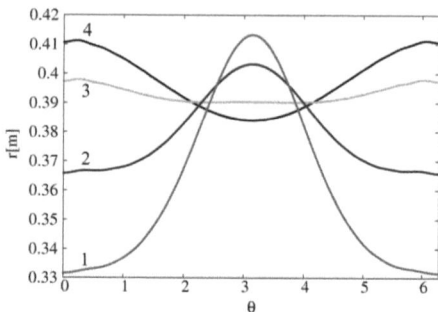

Figure 4.2: Drift surfaces of co-passing electrons for various energies: 2 MeV (line 3), 8 MeV (line 2) and 10 MeV (line 1), with the magnetic KAM surface of the field lines (line 4). All at the irrational value $q = \pi$ of the safety factor.

Figure 4.3: Same as Fig. 4.2 for counter-passing electrons. But here line 1 corresponds to 15 MeV. We used profile "b", given in Sec. 3.3, with $a = 0.46$ m, $R_a = R_0 = 1.75$ m, $I_p = 330$ kA, $B_0 = 2.2$ T, $\beta_{pol} = 0.3$ and $l_i = 1.2$

polar plot, corresponding to the poloidal angle $\theta = \pi$.

Figure 4.3 shows the same as Fig. 4.2 but now for counter-passing electrons. The counter-passing electrons are shifted inwards and again the shift is larger for higher energies. Because the counter-passing particles are drifting inwards, the Shafranov shift is compensated at an energy of about 2 MeV. Figure 4.5 also shows the surfaces for counter passing electrons in a polar plot. As one can see from the polar plots only, the surfaces are not only shifted along the horizontal axis, they are also deformed along the vertical axis. In the case of counter passing particles, see Fig. 4.5, the drift surfaces are stretched along the horizontal axis, while shifted inwards, which causes a deformation from the top and the bottom towards the center. For co-passing electrons it is the other way round, they are deformed from the center towards the top and bottom of the torus, while shifted outwards, according to Fig. 4.4.

The figures of the drift surfaces are all created by the mapping code, but there is another more analytical way to calculate the unperturbed drift surfaces. We will derive an ordinary differential equation for the unperturbed drift surfaces in the (θ, r)-plane from the Hamiltonian equations of motion (4.3.18)-(4.3.23). We assume that there is no electric field, i.e. $\phi = 0$, and that we have no perturbation from the DED, i.e. $f_\varphi = f_\varphi^{(0)}$. We do not need Eqs. (4.3.20) and (4.3.22), because there is no explicit time dependency, and we concentrate on the guiding center. Then we get form Eqs. (4.3.21) and (4.3.23)

$$p_t = \text{const} \quad \text{and} \quad I_x = \text{const} . \tag{4.8.1}$$

4.8 Unperturbed drift surfaces with varying kinetic energy

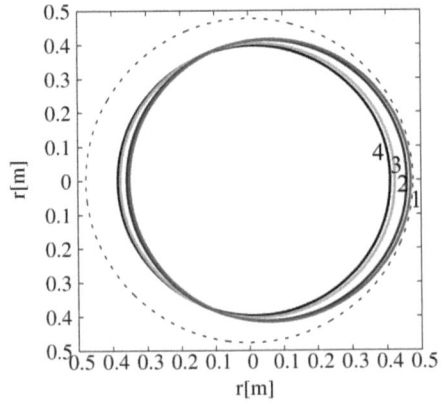
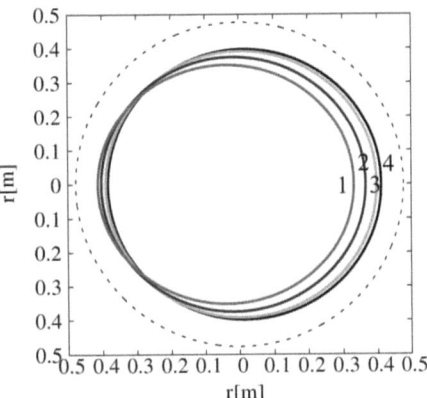

Figure 4.4: Same as Fig. 4.2 plotted in polar coordinates. The tokamak wall is shown as dashed line.

Figure 4.5: Same as Fig. 4.3 plotted in polar coordinates similar to Fig. 4.4

Therefore, we only need Eqs. (4.3.18) and (4.3.19):

$$\frac{dz}{d\varphi} = \frac{1}{Z_q}(1+x_c)\left[\frac{\partial f_\varphi}{\partial x_c} + \sigma\left(\varepsilon_0(\gamma^2-1) - \omega_x I_x\right)\right.$$
$$\left. \times \left(\varepsilon_0(\gamma^2-1) - 2\omega_x I_x\right)^{-1/2}\right] \quad (4.8.2)$$

$$\frac{dp_z}{d\varphi} = \frac{\partial f_\varphi}{\partial z}. \quad (4.8.3)$$

In Eq. (4.8.3) we change the dependent variable p_z to x_c by using Eq. (4.4.12), and get

$$\frac{dx_c}{d\varphi} = \frac{dx_c}{dp_z}\frac{dp_z}{d\varphi} = -\frac{1}{Z_q}(x_c+1)\frac{\partial f_\varphi}{\partial z} = -\frac{1}{q(\rho)g(\rho)}(x_c+1)z. \quad (4.8.4)$$

With Eq. (4.4.11) we find from Eq. (4.8.2)

$$\frac{dz}{d\varphi} = \frac{1}{Z_q}(1+x_c)\frac{\partial f_\varphi}{\partial x_c} + \frac{\sigma}{Z_q}C(x_c) \quad (4.8.5)$$

$$= \frac{1}{q(\rho)g(\rho)}(1+x_c-R_a-\Delta)(1+x_c) + \frac{\sigma}{Z_q}C(x_c), \quad (4.8.6)$$

using the definitions

$$g(\rho) = (1+x_c-R_a+\Delta)\frac{1}{\rho}\frac{\partial \Delta}{\partial \rho} + 1 \quad (4.8.7)$$

and

$$C(x_c) = (x_c+1)\frac{\varepsilon_0(\gamma^2-1) - \omega_x I_x}{\sqrt{\varepsilon_0(\gamma^2-1) - 2\omega_x I_x}}. \quad (4.8.8)$$

Chapter 4. Toroidal DED model with relativistic particle drift effects

We transform to polar coordinates $(x_c, z) \to (r, \theta)$ with

$$r = \sqrt{x_c^2 + z^2} \quad \text{and} \quad \theta = \arctan\frac{z}{x_c}, \quad \text{as well as} \quad x_c = r\cos\theta \quad \text{and} \quad z = r\sin\theta. \quad (4.8.9)$$

Using the derivatives

$$\frac{\partial r}{\partial x_c} = \frac{x_c}{r}, \quad \frac{\partial r}{\partial z} = \frac{z}{r}, \quad \frac{\partial \theta}{\partial x_c} = -\frac{z}{r^2}, \quad \frac{\partial \theta}{\partial z} = \frac{x_c}{r^2}, \quad (4.8.10)$$

we obtain

$$\frac{dr}{d\varphi} = \frac{\partial r}{\partial x_c}\frac{dx_c}{d\varphi} + \frac{\partial r}{\partial z}\frac{dz}{d\varphi} \quad (4.8.11)$$

$$= -\frac{x_c}{r}\frac{1}{q(\rho)g(\rho)}(x_c+1)z + \frac{z}{r}\frac{1}{q(\rho)g(\rho)}(1 + x_c - R_a - \Delta)(1 + x_c)$$

$$+ \frac{\sigma}{Z_q}C(x_c)\frac{z}{r} \quad (4.8.12)$$

$$= \frac{1 + x_c}{q(\rho)g(\rho)r}z(1 - R_a - \Delta) + \frac{\sigma}{Z_q}C(x_c)\frac{z}{r} \quad (4.8.13)$$

and

$$\frac{d\theta}{d\varphi} = \frac{\partial\theta}{\partial x_c}\frac{dx_c}{d\varphi} + \frac{\partial\theta}{\partial z}\frac{dz}{d\varphi} \quad (4.8.14)$$

$$= \frac{z}{r^2}\frac{1}{q(\rho)g(\rho)}(x_c+1)z + \frac{x_c}{r^2}\frac{1}{q(\rho)g(\rho)}(1+x_c-R_a-\Delta)(1+x_c)$$

$$+\frac{\sigma}{Z_q}C(x_c)\frac{x_c}{r^2} \quad (4.8.15)$$

$$= \frac{1+x_c}{q(\rho)g(\rho)r^2}(r^2 + x_c(1 - R_a - \Delta)) + \frac{\sigma}{Z_q}C(x_c)\frac{x_c}{r^2}. \quad (4.8.16)$$

These two differential equations fully describe the unperturbed guiding center motion in the tokamak. In the next step we reduce the description to the (θ, r)-plane, loosing one degree of freedom. The toroidal angle coordinate φ, which is not explicitly used in Eqs. (4.8.13) and (4.8.16), is excluded. Due to the loss of one degree of freedom, the system can no longer show chaotic behavior, as well as islands can no longer appear. But the shape of stable unperturbed drift surfaces remains in the Poincaré plane (θ, r). These drift surfaces are then given

4.8 Unperturbed drift surfaces with varying kinetic energy

by the ordinary differential equation

$$\frac{dr}{d\theta} = \frac{dr}{d\varphi}\frac{d\varphi}{d\theta} \qquad (4.8.17)$$

$$= \frac{z}{r}\frac{1}{q(\rho)g(\rho)}\left[(1+x_c)(1-R_a-\Delta) + \frac{\sigma}{Z_q}q(\rho)g(\rho)C(x_c)\right]$$

$$\times \left\{\frac{x_c}{r^2}\frac{1}{q(\rho)g(\rho)}\left[(1+x_c)\left(\frac{r^2}{x_c}+1-R_a-\Delta\right) + \frac{\sigma}{Z_q}q(\rho)g(\rho)C(x_c)\right]\right\}^{-1} \qquad (4.8.18)$$

$$= \frac{zr}{x_c}\frac{D(x_c,\rho)}{(1+x_c)\frac{r^2}{x_c}+D(x_c,\rho)} . \qquad (4.8.19)$$

Replacing x and z we finally find

$$\frac{dr}{d\theta} = \frac{\sin\theta D(r,\theta)}{1+r\cos\theta+r^{-1}\cos\theta D(r,\theta)} \qquad (4.8.20)$$

with

$$D(r,\theta) = (1+r\cos\theta)(1-R_a-\Delta(\rho)) + \frac{\sigma}{Z_q}q(\rho)g(\rho)C(r\cos\theta) . \qquad (4.8.21)$$

In this equation $\rho = \rho(x_c,z) = \rho(r\cos\theta, r\sin\theta)$ is given by Eq. (4.4.6), and $q(\rho)$ is the safety factor.

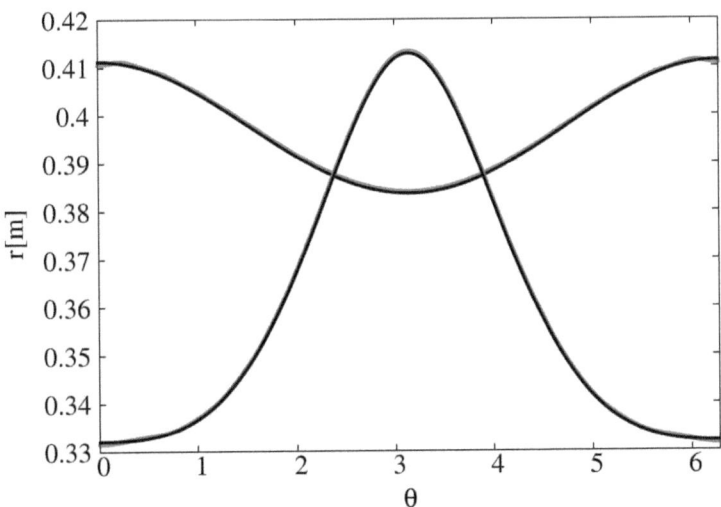

Figure 4.6: Comparison of drift surfaces, calculated by the mapping code, grey lines, and by numerical integration of Eq. (4.8.20), black lines, for field lines and 15 MeV counter-passing electrons. Parameters as above.

We can solve Eq. (4.8.20) with a Runge-Kutta integrator. Figure 4.6 shows the comparison of the results from the mapping code and the Runge-Kutta integration. As an example, we have chosen the irrational $q = \pi$ drift surface for field lines and for 15 MeV counter-passing electrons. As one can see, the results are in excellent agreement. In contrast to the direct numerical integration of the Hamiltonian equations of motion (4.3.18)-(4.3.23) or the mapping code, the Runge-Kutta integration of Eq. (4.8.20) is much faster, because of the reduction to the (θ, r)-plane. The time consuming integration or iteration in φ direction around the torus is not needed here. But note that this simplification is only possible while chaos is excluded.

We can draw some more conclusions from Eq. (4.8.20). The drift effects are described by the term in Eq. (4.8.21) which is proportional to σ. This term is also proportional to $1/Z_q$, i.e. a change of direction corresponds to a change of the sign of the particle charge. So, co-passing electrons are showing the same drift effects than counter-passing positrons for example. This is valid, as long as we do not include an electric field. If we neglect the Shafranov shift, $\Delta(\rho) \equiv 0$ and $R_a = R_0$, we would get results for the cylindrical model and find $dr/d\theta = 0$ for field lines ($\sigma = 0$).

4.9 Drift effects with the DED perturbation field

Now we include the perturbation field of the DED. Using the same parameters as for the unperturbed case, we operate the DED with the perturbation current $I_0 = 10$ kA. In the following we keep this perturbation current constant and vary only the energy of the particles. The mapping procedure (4.7.22)-(4.7.24) provides us with the Poincaré plot, which will be called drift map in the following.

Figure 4.7 shows the drift map for field lines, given by $\sigma = 0$. The structures are a combination of the Shafranov shift, known from Sec. 4.8, and the well known properties of the DED map, see Sec. 3.5. We see four island chains in the ergodic zone, from 12/4 to 15/4, and the finger structures of the laminar zone. The interesting question is now, how do these structures change for drifting particles with different energies. Figures 4.8-4.19 show the drift map for co- and counter-passing electrons on the left and right side, respectively, for energies of 500 keV to 15 MeV.

The structures created by 500 keV co- and counter-passing electrons, shown in Figs. 4.8 and 4.9 respectively, are very similar to the structures of the field lines, see Fig. 4.7. This underlines our assumption that low energy particles are mainly following the field lines. Therefore, the analysis of the field line structures and the field line dynamics is still a very important task. But a plasma does not only consist of low energy particles, which makes it necessary to discuss the differences between the high energy particle dynamics and the field line dynamics.

For co-passing electrons the drift map becomes more and more regular with increasing

4.9 Drift effects with the DED perturbation field

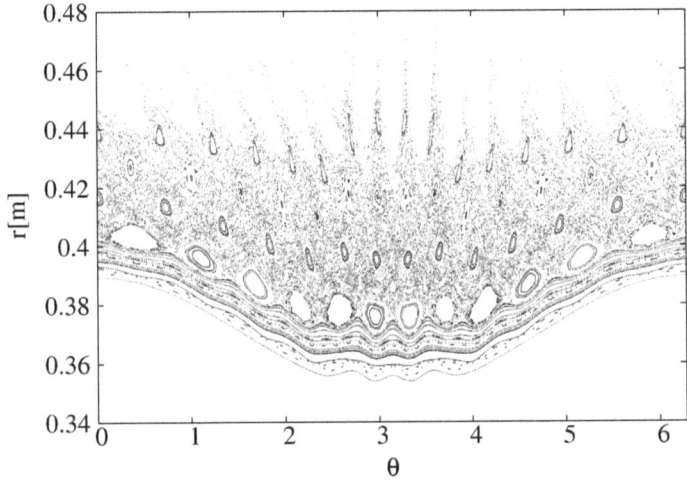

Figure 4.7: Drift map for field lines ($\sigma = 0$) with $I_0 = 10$ kA

particle energy. Although the structures in Fig. 4.10 are still similar to the field line structures, the map has already become more regular. For example the 12/4 island chain is no longer inside the ergodic zone. Also the finger structure is less pronounced as for field lines. At a particle energy of 5 MeV the ergodic zone has nearly vanished, while the laminar zone is still visible, as Fig. 4.12 shows. As one can see from e.g. Fig. 4.14, the finger-structures of the laminar zone have vanished along with the ergodic zone. The co-passing electrons with a kinetic energy of 8 MeV and more are confined inside the plasma. The intact drift surfaces are now connected to the tokamak wall at the outside of the torus, corresponding to the poloidal angle $\theta = 0$. In Fig. 4.18 the electrons behave, as if there would be no perturbation at all, except of a few very small islands. Due to the drift effects, the structures of co-passing electrons are shifted outside, so that the particles are shifted away from the DED coils, as already shown in Sec. 4.8. They are shifted into areas of lower perturbation. Therefore, it is clear that the drift map becomes more and more regular for higher energies and also the vanishing of the finger structures becomes clear.

But there is another effect, which also contributes to the fact that the high energetic electrons behave more regular. This can be seen from the counter-passing electrons. The drift map for counter-passing electrons becomes also more and more regular with increasing particle energy, but the finger-structures become more distinctive and much more concentrated at the angle position of the DED coils, $\theta = \pi$, according to e.g. Fig. 4.13. Due to the drift effects, which cause a shift of the drift surfaces towards the DED coils, as known from Sec. 4.8, the occurrence

Chapter 4. Toroidal DED model with relativistic particle drift effects

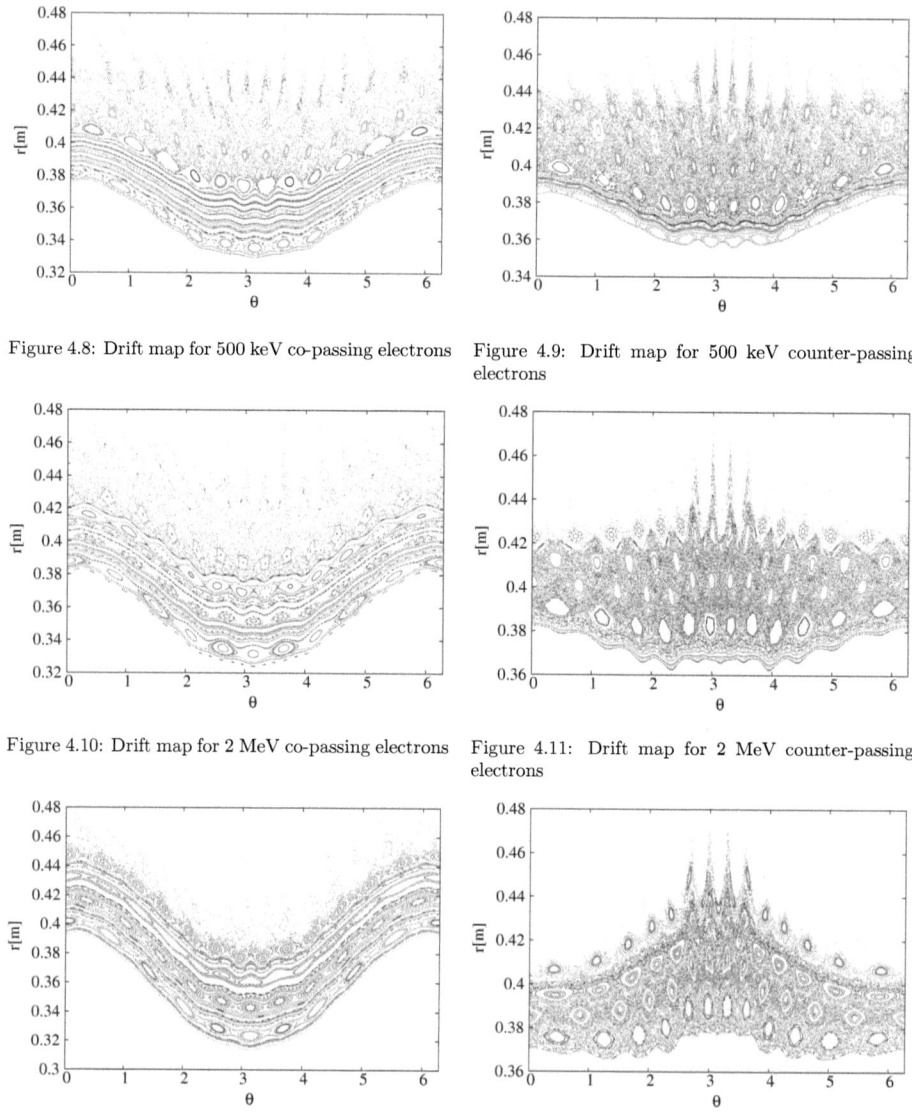

Figure 4.8: Drift map for 500 keV co-passing electrons

Figure 4.9: Drift map for 500 keV counter-passing electrons

Figure 4.10: Drift map for 2 MeV co-passing electrons

Figure 4.11: Drift map for 2 MeV counter-passing electrons

Figure 4.12: Drift map for 5 MeV co-passing electrons

Figure 4.13: Drift map for 5 MeV counter-passing electrons

4.9 Drift effects with the DED perturbation field

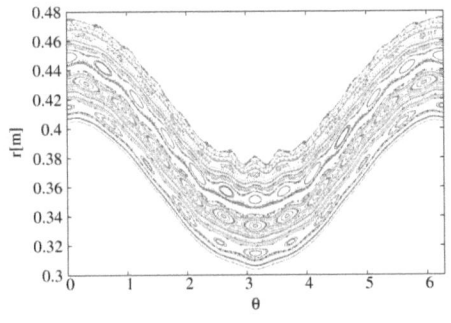

Figure 4.14: Drift map for 8 MeV co-passing electrons

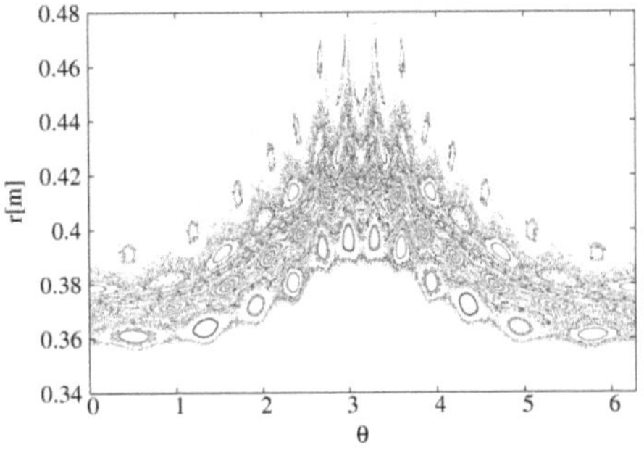

Figure 4.15: Drift map for 8 MeV counter-passing electrons

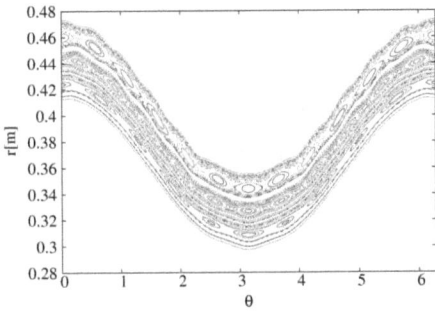
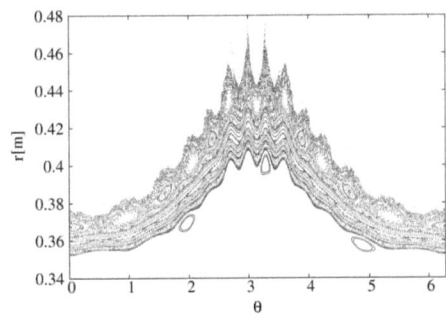

Figure 4.16: Drift map for 10 MeV co-passing electrons

Figure 4.17: Drift map for 10 MeV counter-passing electrons

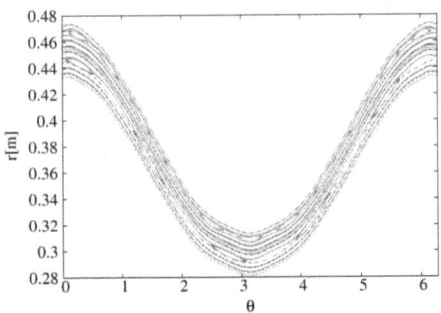
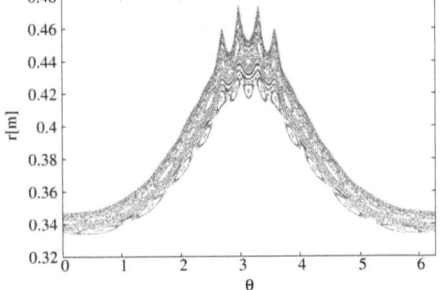

Figure 4.18: Drift map for 15 MeV co-passing electrons

Figure 4.19: Drift map for 15 MeV counter-passing electrons

of the fingers around $\theta = \pi$ is expected. But the counter passing electrons are shifted into areas of stronger perturbations. One would expect similar or even stronger chaotic behavior. In the contrary, the counter passing electrons also behave more regular for higher energies, as can be seen in Figs. 4.17 and 4.19. This can only be an effect of the high kinetic energy of the electrons. Their motion inside the torus is so fast that they can not really feel the chaotic magnetic field. Some effects of the perturbation are still present, according to Fig. 4.19, where we can clearly see four fingers at the position of the coils. But also, as for co-passing electrons, intact drift surfaces are connected to the wall, now at the inner side of the torus.

We can conclude that generally particles with high kinetic energy do not really feel the chaotic magnetic field. The high energetic particles are confined inside the plasma, and they are strongly shifted to the inside or the outside of the torus for counter- or co-passing electrons, respectively. The last intact drift surface is located directly beneath the wall either at the outside, $\theta = 0$, or at the inside, $\theta = \pi$, of the torus, depending on the particles and their direction of motion. Particles beyond the last intact drift surface are connected to the wall, because they move on surfaces, which cross the wall. These particles are lost at the wall extremely fast.

4.10 Escape rates of particles and field lines

In order to quantify the result that the behavior of particles become more regular with increasing kinetic energy, we calculate the escape rates of, e.g., counter-passing electrons for different energies. We choose N_0 "test particles" on the irrational $q = \pi$ surface, equally distributed along the whole poloidal angle range, and iterate them until they hit the wall, where they are eliminated. The calculation is stopped, when 90% of the points are lost.

Figure 4.20 shows the calculated escape rates for field lines, 1.25 MeV and 1.75 MeV counter-passing electrons. The perturbation current is kept constant at $I_0 = 10$ kA. As one can see, the escape rates are in very good agreement with the exponential decay fitting curves

$$N(t) = N_0(N_1 + e^{-\lambda t}) . \tag{4.10.1}$$

The very good agreement is underlined by Fig. 4.21, where the logarithms

$$\ln\left(\frac{N}{N_0} - N_1\right) = -\lambda t \tag{4.10.2}$$

are shown. The logarithms of the escape rates are in excellent agreement with the linear fits, given by the grey lines of Fig. 4.21. From here we can precisely determine the decay parameter λ, given by the gradient of the linear fits. Note that there is an offset N_1 of about 5-10% for all escape rates. The reason for this offset is unknown. One can speculate that these extremely

4.10 Escape rates of particles and field lines

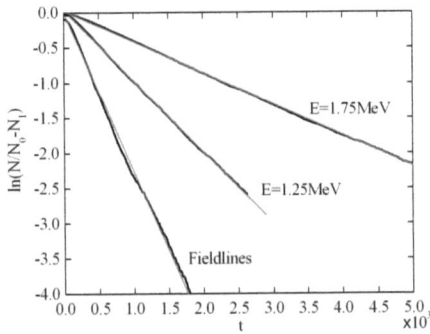
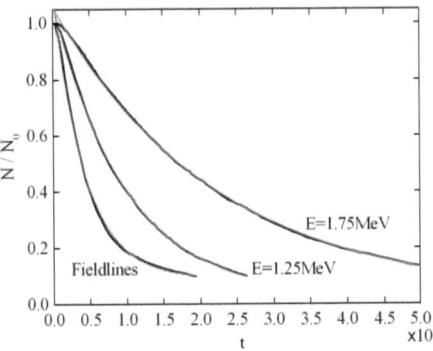

Figure 4.20: Normalized escape rates of counter-passing electrons, black lines, with exponential decay fitting curves, grey lines. Calculated with $N_0 = 5000$ at a constant perturbation of $I_0 = 10$ kA.

Figure 4.21: Logarithm of the escape rates of Fig. 4.20, black lines, with linear fits, grey lines.

long running particles are either following another, very long time escape mechanism or that they are not leaving the system at all, because they are moving on the stable manifold for example. By calculating the escape rates for various kinetic energies, we find the dependency of the characteristical decay parameter λ on the energy as follows

$$\lambda \sim -(E - E_{crit}) \,. \tag{4.10.3}$$

This is shown in Fig. 4.22. The decay parameter decreases linearly with increasing energy, which clearly approves the conclusions drawn from the Poincaré plots. From Fig. 4.22 we can extrapolate the critical energy E_{crit}, where a stable drift surface is formed at the edge of the ergodic zone, which confines all particles of the ergodic zone inside the plasma. We obtain

$$\lambda = 0 \quad \Leftrightarrow \quad E_{crit} = 2.176 \text{ MeV} \,. \tag{4.10.4}$$

The laminar zone is still open to the wall. From the Poincaré plots one can see that the stable drift surface is formed between the 14/4 and 15/4 island chains, while the 15/4 island chain is the last remaining resonance at the transition of the ergodic to the laminar zone.

This stabilization effect is compatible with the stabilization due to a reduction of the perturbation current I_0. Figure 4.23 shows the dependency of the square root of λ on I_0. Here we take only field lines into account. The escape rates of the field lines for various perturbations follow the same exponential decay law (4.10.1) as the escape rates of particles for various energies, but

$$\lambda \sim (I_0 - I_{crit})^2 \tag{4.10.5}$$

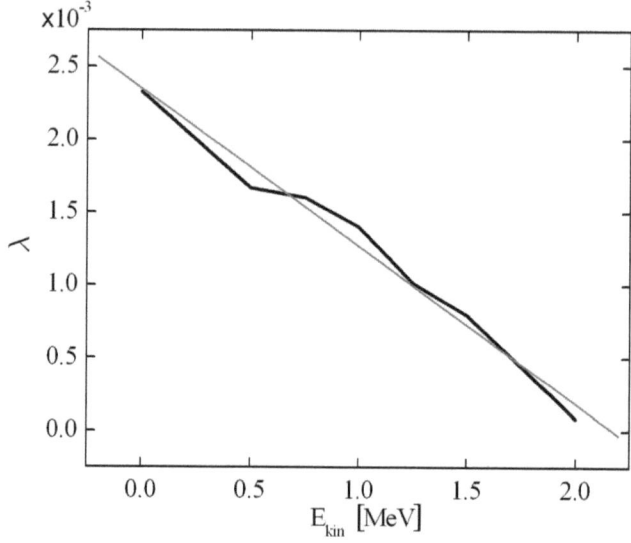

Figure 4.22: Exponential decay parameter λ in dependency of the kinetic energy, black line, with linear fit, grey line.

with the critical perturbation current I_{crit}, where a stable KAM surface is formed at the edge of the ergodic zone. By extrapolation we get

$$\lambda = 0 \quad \Leftrightarrow \quad I_{crit} = 7.75 \text{ kA} . \qquad (4.10.6)$$

The energy of the magnetic perturbation field scales with the square of the perturbation current

$$E_0 = \frac{1}{2} L I_0^2 , \qquad (4.10.7)$$

while L is the sum of the inductances of the DED coils. Therefore, the stabilization effect of decreasing perturbation is of the same order as the effect of increasing kinetic energy and scales linearly with the corresponding energy. On the contrary to the tokamap, where the characteristical decay parameter (2.4.7) for the escape rate scales with the third power of the perturbation parameter, the decay parameter of the drift map scales only with the square of the perturbation. Note, in both cases the perturbation part of the Hamiltonian scales linearly with the perturbation parameter.

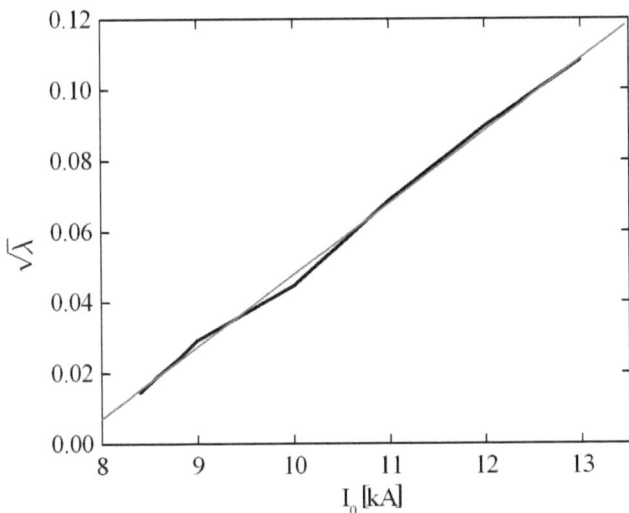

Figure 4.23: Square root of the decay parameter λ in dependency of the perturbation current I_0 for field lines, black line, with linear fit, grey line.

4.11 Non-relativistic Limit

The kinetic energies of the particles in the plasma are given by the Maxwell distribution, which means that the probability p to find a particle with the kinetic energy E obeys

$$p \sim e^{-E/kT}, \qquad (4.11.1)$$

while T is the temperature of the particles. Therefore, only some particles of the plasma have energies equal or larger than the rest energy at the typical fusion plasma temperatures of $kT = 10 - 20$ keV at the plasma center. Most of the particles have energies much less than the rest energy so that they can be described non-relativistically. Because the relativistic description, which is derived in the first three sections here, includes the non-relativistic one as a special case, we can go to the $E \ll \varepsilon_0$ limit to obtain the non-relativistic Hamiltonian and equations of motion.

The relativistic Hamiltonian for the particle drift, using the toroidal angle as independent variable, is given by Eq. (4.3.17) and reads

$$K = -f_\varphi - \sigma(1 + x_c)\left[\varepsilon_0(\gamma^2 - 1) - 2\omega_x I_x\right]^{1/2} \qquad (4.11.2)$$

Chapter 4. Toroidal DED model with relativistic particle drift effects

with the normalized rest energy

$$\varepsilon_0 = \frac{c^2}{\omega_c^2 R_0^2} \tag{4.11.3}$$

and

$$\gamma = \frac{-p_t - \phi}{\varepsilon_0} . \tag{4.11.4}$$

The term

$$\varepsilon_0(\gamma^2 - 1) = \varepsilon_0 \left(\frac{-p_t - \phi}{\varepsilon_0}\right)^2 - \varepsilon_0 \tag{4.11.5}$$

is the only part of the Hamiltonian K, which depends on the normalized total particle energy

$$\tilde{H} = -p_t . \tag{4.11.6}$$

For the non-relativistic limit, we have to separate the non-relativistic part of the energy from the total relativistic energy by subtracting the rest energy

$$\tilde{H}_{nr} = \tilde{H} - \varepsilon_0 = -p_t - \varepsilon_0 =: -h . \tag{4.11.7}$$

Here we introduced the canonical momentum h, which is used for the non-relativistic case. It corresponds to the negative non-relativistic total energy. Inserting h into Eq. (4.11.5), we obtain

$$\varepsilon_0 \left(\frac{-p_t - \phi}{\varepsilon_0}\right)^2 - \varepsilon_0 = \varepsilon_0 \left(\frac{-h + \varepsilon_0 - \phi}{\varepsilon_0}\right)^2 - \varepsilon_0 = \varepsilon_0 \left[\left(1 + \frac{-h - \phi}{\varepsilon_0}\right)^2 - 1\right] . \tag{4.11.8}$$

Going to the non-relativistic limit now means that the kinetic energy, given by the difference of total energy h and electric potential ϕ, is much smaller than the rest energy ε_0. Therefore, we introduce the smallness parameter

$$\mu := \frac{-h - \phi}{\varepsilon_0} \ll 1 . \tag{4.11.9}$$

Now we can expand the right side of Eq. (4.11.8) into a power series with respect to μ

$$\varepsilon_0 \left[(1 + \mu)^2 - 1\right] \approx 0 + 2\varepsilon_0 \mu + O(\mu^2) = 2(-h - \phi) \tag{4.11.10}$$

and neglect all higher orders of μ. Inserting this result into the relativistic Hamiltonian K, we get the non-relativistic Hamiltonian

$$K_{nr} = -f_\varphi - \sigma(1 + x_c)\sqrt{2(-h - \phi - \omega_x I_x)} \tag{4.11.11}$$

and the non-relativistic equations of motion [20]

$$\frac{dz}{d\varphi} = \frac{1}{Z_q}(1+x_c)\left[\frac{\partial f_\varphi}{\partial x_c} + \sigma\left(2(-h-\phi) - \omega_x I_x - (1+x_c)\frac{\partial \phi}{\partial x_c}\right)\right.$$
$$\left. \times (2(-h - \omega_x I_x - \phi))^{-1/2}\right] \quad (4.11.12)$$

$$\frac{dp_z}{d\varphi} = \frac{\partial f_\varphi}{\partial z} - \sigma(1+x_c)\frac{\partial \phi}{\partial z}(2(-h - \omega_x I_x - \phi))^{-1/2} \quad (4.11.13)$$

$$\frac{dt}{d\varphi} = \sigma(1+x_c)(2(-h - \omega_x I_x - \phi))^{-1/2} \quad (4.11.14)$$

$$\frac{dh}{d\varphi} = \frac{\partial f_\varphi}{\partial t} - \sigma(1+x_c)\frac{\partial \phi}{\partial t}(2(-h - \omega_x I_x - \phi))^{-1/2} \quad (4.11.15)$$

$$\frac{d\vartheta_x}{d\varphi} = \sigma(2(-h - \omega_x I_x - \phi))^{-1/2} \quad (4.11.16)$$

$$\frac{dI_x}{d\varphi} = 0. \quad (4.11.17)$$

In [20] the drift effects of non-relativistic ions were discussed. For ions the drift effects are much larger than for electrons, because of the larger mass of the ions. Also ions can be described non-relativistically at the same energy levels, where electrons are already relativistic, due to the much larger rest energy of ions, e.g. 938.27 MeV for protons. Because the results for the non-relativistic ions are similar to the results shown here for electrons, they are not outlined here. Note, the only difference is that the results on co-passing electrons correspond to the results on counter-passing ions, because of the different sign of charge.

4.12 Heat flux patterns in TEXTOR

After analyzing the dynamics of the chaotic DED system theoretically, we apply our results, especially on the stable and unstable manifolds, to measurements of the heat flux at the divertor plates of the TEXTOR-DED experiment. We will show that these heat flux patterns can be explained and fully understood by the stable and unstable manifolds. Especially the manifolds of the last island chain in the ergodic zone, which is located at the transition to the laminar zone, will be important.

Figure 4.24 shows a measurement of the heat flux pattern in dependency of the edge safety factor. This measurement has been done by Marcin Jakubowski from the Forschungszentrum Jülich. The pattern is taken at a fixed toroidal position over a small poloidal angle area. The figure shows the development of the heat flux with changing edge safety factor

$$q_a = \frac{2\pi B_0 R_0 a^2}{\mu_0 I_p}\left(1 + \frac{1}{2}A_1 a^2 + \frac{3}{8}A_2 a^4\right). \quad (4.12.1)$$

Chapter 4. Toroidal DED model with relativistic particle drift effects

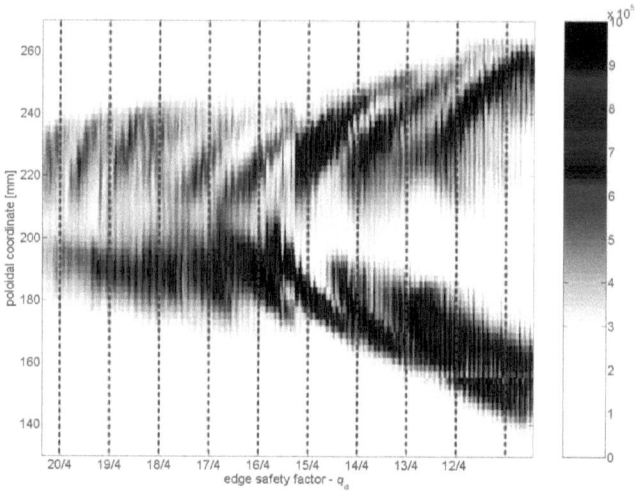

Figure 4.24: Measurement of the heat flux pattern with changing edge safety factor q_a at a fixed toroidal position of the divertor plates. The dark color indicates hot areas on the divertor plate. This measurement has been done by M. Jakubowski at the TEXTOR-DED experiment of the Forschungszentrum Jülich.

The term in the brackets is a toroidal correction to the edge safety factor (3.3.10) of the cylindrical model. The correction parameters are $A_1 = 3 + 2\Lambda + \Lambda^2$ and $A_2 = 5 + 4\Lambda + 3\Lambda^2 + 2\Lambda^3 + \Lambda^4$, which depend on the Shafranov parameter Λ. Note that a is the minor plasma radius, normalized with the major radius R_0 of the torus. The edge safety factor depends essentially on the plasma current I_p, which is varied here to change q_a. All other parameters are kept constant as follows: perturbation current $I_0 = 11.4$ kA, toroidal field $B_0 = 1.93$ T, minor radius of plasma $a = 0.437$ m, major radius of plasma $R_a = 1.7$ m, internal inductance $l_i = 1.2$ and plasma beta $\beta_{pol} = 0.45$. The divertor plates are located at a minor radius of $r_w = 0.477$ m.

The measurement shows that at certain values of q_a new strike zones appear, while the former strike zone tends outwards, compared to the center of the figure, gets smaller and finally vanishes. Important is that the strike zones are overlapping, which means that the next strike zone appears, while the last one has not vanished. These structures of the heat flux pattern can be explained by the theory of the stable and unstable manifolds.

As already mention, most of the plasma particles are moving with velocities close to the thermal velocity

$$v_{th} = \left(\frac{kT}{m}\right)^{1/2}, \qquad (4.12.2)$$

4.12 Heat flux patterns in TEXTOR

given by the plasma temperature of $kT = 10 - 20$ keV at the plasma center. According to our analysis of the drift effects for high energetic particles in Sec. 4.9, we have seen that low energy particles are mainly following the magnetic field lines, while the high energy particles are mainly confined inside the plasma. Therefore, we will analyze the heat flux pattern with the field line dynamics.

In order to create the heat flux pattern of Fig. 4.24 theoretically, we calculated 97 laminar plots for different plasma currents, from $I_p = 228$ kA to $I_p = 420$ kA in 2 kA steps. Connecting the values at the edges of these laminar plots, we obtain Figs. 4.25 and 4.26, using the same parameters as used for the measurement.

According to the definition of a laminar plot, the dark colored areas correspond to the strike points of those field lines with the wall, which have very large connection lengths. Here more than 10 toroidal iterations are needed for field lines of the black areas to get from wall to wall inside the torus. From the analysis of the revtokamap and the cylindrical DED model, we know that field lines which are exactly on the stable and unstable manifolds have infinite connection lengths so that the dark colored areas are directly related to the strike points of the stable and unstable manifolds with the wall. We know further that the field lines are following the unstable manifolds to the wall. Therefore, the dark colored areas are identical to those areas of the wall, where the plasma particles are hitting. Because the manifolds deeply penetrate the plasma, the largest amount of heat deposition has to be at the strike points of the manifolds. But not only the unstable manifolds are important. Although the field lines are following the unstable manifolds to the wall, the plasma particles are either co-passing or counter-passing, which means that they are either moving in the direction of the field lines and therefore are following the unstable manifold, or against so that they are following the stable manifold, which is the unstable one for the reverse direction. Because of this, Fig. 4.25 also shows the heat flux pattern at the wall within the total poloidal angle area $\pi - \theta_c \leq \theta \leq \pi + \theta_c$ of the coils, while in Fig. 4.26 the stripes around $\theta = 3.3$ of Fig. 4.25 are magnified.

Figure 4.26 is in a very good agreement with the effects, we can see in the measurement. We can identify various different strike zones, which are overlapping in the same way, we are observing in the experiment. One can also clearly see the divergence of the two stripes of strike zones with decreasing q_a. Also quantitatively the agreement is very good, according to the numbers of strike zones. In the measurement, we can clearly identify 4 different strike zones in the range of $q_a = 16/4 = 4$ to $q_a = 12/4 = 3$. In the numerical calculation, we also count 4 strike zones in this q_a range. The only slight difference is that the inner branches of the "c"-shaped strike zones cannot be clearly observed in the measurement. But, as can be seen from Fig. 4.26, they are not as distinct as the outer branches. We will come back to this later.

Now we analyze the structures in detail. We shall see that each strike zone is directly related to the last resonance at the edge of the ergodic zone for the respective parameter regime, which

Chapter 4. Toroidal DED model with relativistic particle drift effects

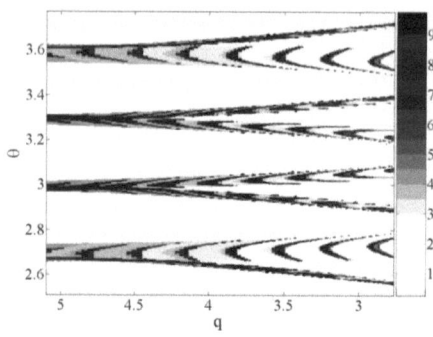
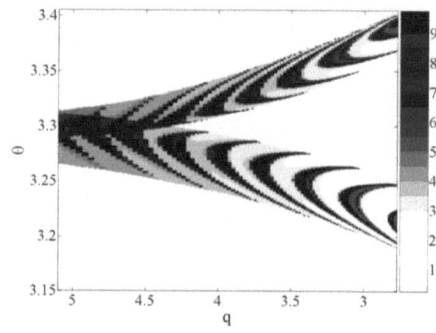

Figure 4.25: Development of the wall pattern in the total angle area $2\theta_c$ of the DED coils at a fixed toroidal position with decreasing edge safety factor q_a.

Figure 4.26: Magnification of the stripe around $\theta = 3.3$ of Fig. 4.25.

is then also the dominant one. Due to the decrease of the edge safety factor, the resonances are shifted towards the wall and therefore, they are destroyed. We have picked out three different cases to prove out statement. Figure 4.27 shows the drift map for field lines for an edge safety factor of $q_a = 3.13$. As one can see, the last resonance is the 10/4 island chain. A stable manifold, given by the grey line, and an unstable one, given by the black line, which hit the wall in the considered angle area of $3.15 \leq \theta \leq 3.4$, are shown. These manifolds are manifolds of period 5 hyperbolic points, which are located in the 10/4 island chain. According to the corresponding laminar plot, Fig. 4.28, one can see that, as already known, the dark colored areas are given by the manifolds. Further, the left strike zone is given by the stable manifolds, while the right one is given by the unstable ones.

Each "c"-shaped strike zone has two branches. Due to the overlap of the strike zones, there are several branches the manifolds can be related to. Using the laminar plots, we can clearly identify which branches are related to the considered manifolds. In Fig. 4.33, which is a magnification of the last three strike zones of Fig. 4.26, the strike points of the stable and unstable manifolds with the wall are shown as white and grey points, respectively. The points indexed with "1" correspond to the strike points of the manifolds of Figs. 4.27 and 4.28. Therefore, we can conclude that these strike zones are related to the 10/4 resonance.

The dominance of the 10/4 resonance can be concluded from Fig. 4.29. Here the drift map is shown for a smaller value of $q_a = 3.03$. For this value of q_a the 10/4 resonance has become very small, but it is still present. The stable and unstable manifolds, which are shown, are manifolds of period 9 hyperbolic points of the 9/4 resonance, located below the 10/4 resonance.

4.12 Heat flux patterns in TEXTOR

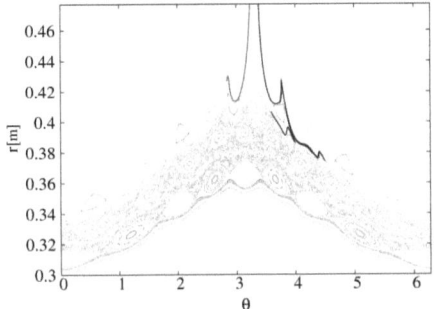
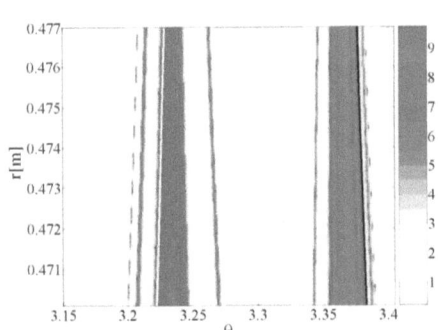

Figure 4.27: Drift map for field lines with $q_a = 3.13$. Stable (grey line) and unstable manifold (black line) for the strike zones of Fig. 4.26 are shown.

Figure 4.28: Laminar plot of the relevant angle area close to the wall, corresponding to Fig. 4.27. The stable (white line) and unstable manifold (black line) are shown.

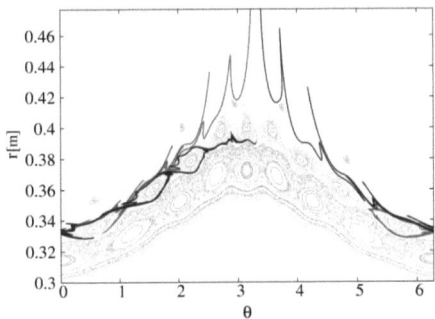
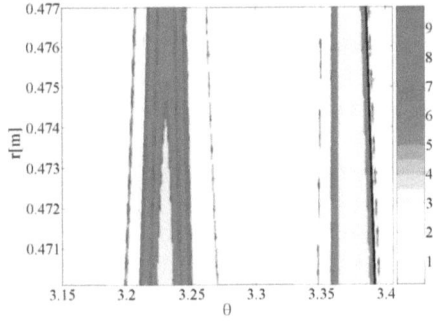

Figure 4.29: Same as Fig. 4.27 but for $q_a = 3.03$.

Figure 4.30: Same as Fig. 4.28 but corresponding to Fig. 4.29

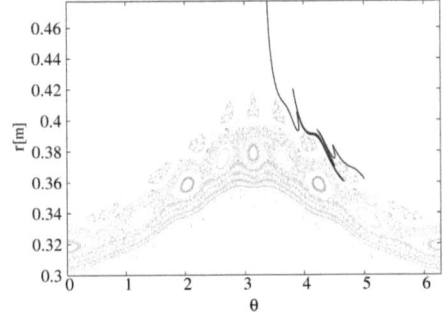
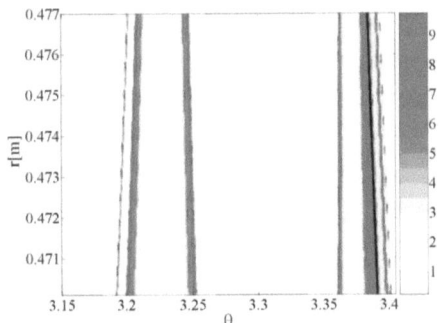

Figure 4.31: Same as Fig. 4.27 but for $q_a = 2.94$.

Figure 4.32: Same as Fig. 4.28 but corresponding to Fig. 4.31

Their strike points at the wall are marked by the white and grey points, respectively, which are indexed by "2" in Fig. 4.33. These strike points are located in the branches, which belong to the 10/4 resonance strike zone. So the 10/4 resonance is still dominant, because the manifolds of the period 9 hyperbolic points pass the 10/4 island chain on their way to the wall. As already mentioned in previous sections, the manifolds then converge towards the manifolds of the last island chain and follow their path towards the wall. This shows that all resonances below the last, dominant one and their manifolds have no direct effect on the wall pattern.

To underline this conclusion, we consider a third case with $q_a = 2.94$. For this case, the drift map is shown in Fig. 4.31. Here we can see that the 10/4 island chain is just destroyed and now the 9/4 island chain is the last one. The shown unstable manifold corresponds to a period 9 hyperbolic point. The laminar plot for this case, figure 4.32, shows that this unstable manifold, given by the black line, rules the main branch of the right strike zone. According to Fig. 4.33, where the strike point of this manifold is given by the grey point, indexed with "3", this branch belongs to the next "c"-shaped strike zone. So we can conclude that this is the strike zone of the 9/4 resonance, which has right now become dominant.

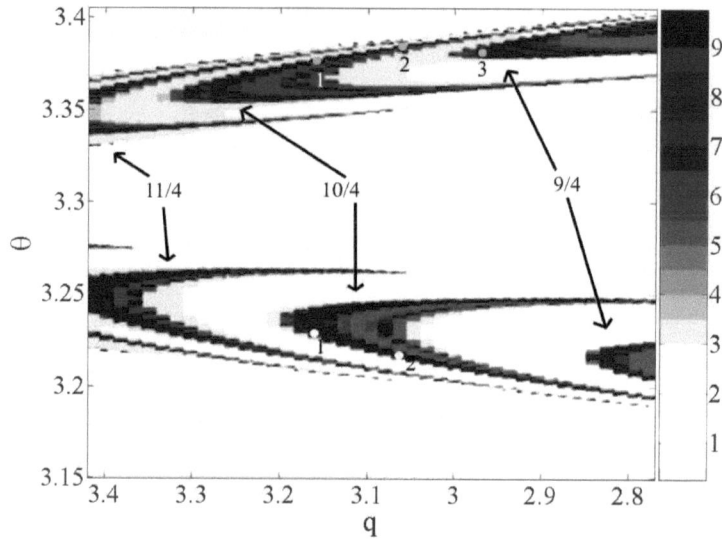

Figure 4.33: Magnification of the last 3 strike zones of Fig. 4.26. The to the strike zones corresponding resonances are marked. The strike points of several stable (white points) and unstable manifolds (grey points) are shown.

Figure 4.33 combines all results of the detailed structure analysis. Each strike zone is related to one resonance of the Poincaré section. This resonance is the last remaining island chain at the edge of the ergodic zone for the respective q_a value. This resonance dominates all other resonances below, because to their way to the wall, the field lines have to pass this last resonance. Decreasing the edge safety factor shifts the resonances towards the wall, whereas they are destroyed. Therefore, at a certain point, when this island chain is destroyed, the next resonance becomes dominant, which results in a new strike zone. Which strike zone belongs to which resonance can be determined by the analysis of the stable and unstable manifolds and is shown in Fig. 4.33. When the next resonance becomes dominant, most of the particles hit the wall at the corresponding strike zone. The effects of the previously dominant resonance, although it can no longer be observed in the Poincaré plot, have not completely vanished. Because the dominant resonance can only rule the effects of itself and the resonances below, the effects of resonances above are not affected. This explains the overlapping of the strike zones. Even if the main effects are given by the dominant resonance and even if the resonances above can no longer be observed in the Poincaré plot, there are still some small influences of them. These small influences vanish not until the value of q_a has reached the q-value of the respective resonance.

According to the analysis of the stable and unstable manifolds, the strike zones of one stripe are all related to either the stable or the unstable manifolds. For Figs. 4.26 and 4.33 this means that the upper stripe is created by the unstable manifolds, while the lower one is created by the stable manifolds. Then for the four dual stripe structures in Fig. 4.25 the same conclusion is valid. As one can easily see, Fig. 4.25 is symmetrical to $\theta = \pi$, but the dual stripe structures themselves are not symmetrical. According to Fig. 4.33 the strike zones, related to the same resonance, appear at smaller values of q_a the closer the stripes are to the symmetry axis $\theta = \pi$. This effect is caused by the Shafranov shift of the plasma towards the coils.

As already mentioned, the inner branches of the strike zones, according to Figs. 4.26 and 4.33, cannot be clearly observed in the measurement. As it can be seen from the measurement, the outer branches are much more distinct than the inner ones. This can be anticipated also from the figures. For the detailed analysis of the structures, the manifolds have been calculated. The numerical calculation of the various manifolds of one island chain depends on certain initial conditions, e.g. for the Newton method of the mapping procedure, and parameters like the step-size. This parameter regime is not similar to all manifolds. Most of the manifolds can be calculated easily with a broad spectrum of parameters, but other manifolds can only be calculated with very specific combinations of parameters. Due to this fact, not all manifolds of the considered island chains have been found. The ones which habe been found and are shown here, all strike at the outer branches, which are the dominant parts of the strike zones, according to the measurements. This indicates that the problem of the manifold calculation is not only numerical. Different manifolds have different basins of attraction, which means that some manifolds are attracting a large amount of the particles, while other manifolds only attract few. These dominant manifolds are the ones striking at the outer branches. This explains the strong dominance of the outer branches and the numerical problems of calculating all manifolds. The laminar plot on the other hand shows all effects of all manifolds, and it is difficult to decide, which structures of the laminar plot are the dominant ones. Only the combination of laminar plots with stable and unstable manifolds reveals the dominant effects and structures. The combination of both is then in perfect agreement with the measurements.

Chapter 5

Summary and conclusion

Using the Hamiltonian mapping technique, presented in Sec. 2.2, we analyzed the wall patterns and transport mechanisms in open chaotic systems. For this purpose we used the stable and unstable manifolds of hyperbolic periodic points in particular.

First we studied a basic model. For this, we derived the symmetric tokamap and compared it with the non-symmetric one, originally proposed by Balescu, by analyzing their statistical properties. We found that there are some important qualitative and quantitative differences between both maps. The non-symmetric tokamap cannot be constructed from a similar continuous Hamiltonian system under the constraints of the Hamiltonian and generating function as discussed in Sec. 2.1.

The quasi-linear diffusion was analyzed for very large perturbation parameters. Both maps show the same diffusive behavior and the same dependency of the diffusion coefficient on the perturbation parameter ε. The diffusion coefficient is proportional to ε^2. Compared to the quasi-linear diffusion coefficient of the standard map, the q-profile is responsible for a different behavior of the diffusion coefficient around its mean value (the quasi-linear limit). The quasi-linear diffusion coefficient of the tokamap is smaller than the one of the standard map. This can be explained by the density distribution of the intersection points of the field lines with the (ψ, ϑ)-plane. For very large perturbations, $\varepsilon \gg 1$, the points are equally distributed on the plane, but very close to $\psi = 0$ the density increases extremely; it even tends to infinity at $\psi = 0$.

The critical perturbations for the break-up of the two last intact KAM surfaces of both maps were determined. The symmetric tokamap is much more robust against perturbations than the non-symmetric one. Qualitatively, this can be recognized from Figs. 2.1 and 2.2. Quantitatively, this impression was confirmed by determining the critical perturbations for the break-up of KAM surfaces. The critical parameters are significantly larger for the symmetric tokamap than for the non-symmetric one. More important is that the order for the break-up of the two surfaces is different in both maps. For the anomalous transport behavior through

Chapter 5. Summary and conclusion

the noble KAM tori [29] we derived the dependency of the escape rates on the perturbation parameter for the symmetric tokamap. The number of "test-particles" below the broken KAM surface decays exponentially, while the characteristic parameter scales with the third power of ε.

Typical structures of stable and unstable manifolds were presented in Figs. 2.6 and 2.7. The appearance of chaos around the hyperbolic points was explained by the structures and the interactions of the stable and unstable manifolds. The hyperbolic points are the source of chaotic motion. Due to the slightest perturbation the ideal separatrix splits into the stable and unstable manifold, which show a strong oscillatoric behavior close to the hyperbolic points. Due to the intersection of the loops of the stable manifold with the loops of the unstable one, a fractal structure is formed, which causes chaotic motion around the hyperbolic points.

Transport between different stochastic layers was explained by the overlap of the manifolds of neighboring island chains, shown in Fig. 2.8. The transport mechanism was described and is sketched in Fig. 2.9. The break up of intact KAM surfaces and the anomalous transport through the just broken KAM surface are caused by intersections of the manifolds. Therefore, the manifolds are playing a fundamental role in the formation of stochastic layers and chaotic transport mechanisms.

The proposed [29] spontaneous inversion of the q-profile due to perturbations does not exist. However, when already the zeroth-order q-profile shows a reverse-shear behavior, the so called symmetric revtokamap can be derived. The symplectic symmetric revtokamap differs considerably from the previously proposed non-symmetric revtokamap. The open chaotic system can be characterized by the laminar plots, which have been introduced in this context. The laminar plots characterize basins of field line connection lengthes for wall to wall connections inside the plasma. The topological structure of the laminar plots can be understood from the geometrical behavior of the stable and unstable manifolds of the last island chain. It is a fractal one. Dividing the structures of the laminar plot into the structures created by forward or backward iteration only, we have seen that the stable and unstable manifolds of the last island chain determine the border lines between the basins. This indicates that the manifolds of the hyperbolic points of the last island chain are responsible for the transport to the wall. Their topological structures determine the so called footprints at, e.g., the divertor plates. These results are published in [18].

The results, discussed for the tokamap and the revtokamap, are typical for the magnetic field line behavior in stochastic plasmas. The methods proposed can be used to investigate various applications, e.g. the DED at TEXTOR. Therefore, we considered the cylindrical DED model to apply the previous results to a more realistic model. Although the real TEXTOR-DED experiment of the Forschungszentrum Jülich has a toroidal geometry, we started with a cylindrical model to concentrate on the effects of the DED itself. Due to the toroidal geometry several corrections would have to be applied to, e.g., the safety factor and the perturbation spectrum. The most important advantage of the cylindrical model is that it can be calculated totally analytical.

We calculated the magnetic field of the DED coil system, regarding the geometry and a considered current distribution. Then the Hamiltonian was derived, using the Clebsch representation, and finally in Sec. 3.5 the DED map was constructed in the symmetrical form, according to the mapping technique. The typical structure of the DED map can be seen in Fig. 3.4. To classify the open chaotic DED system we analyzed its statistical properties. The DED system shows subdiffusive transport behavior, which was concluded from the MSD. For long time iterations, the MSD decreases due to the loss of field lines. Only the ones, sticking around the islands, remain. To characterize the stochastic motion within the chaotic layers, the Lyapunov exponent was calculated in Sec. 3.6. It shows no dependency on the poloidal angle, while the mean value depends on the radius. Using the mean Lyapunov exponent the Kolmogorov length has been introduced. According to the variation of the Kolmogorov length with the radius, the ergodic and laminar zones were specified.

The typical structures of the stable and unstable manifolds of the DED map were presented in Fig. 3.9. The analysis of the manifolds clearly showed that the field lines are following the unstable manifolds to the wall. The sticking around the islands is also related to the manifolds. The finger-like structures of the DED map are ruled by the unstable manifolds, especially the ones of the last island chain. This underlines the conclusions, drawn from the revtokamap, that the heat and particle transport to the wall and the wall patterns are primarily created by the stable and unstable manifolds of the last resonance.

Also the structures within the laminar zone were analyzed, using the laminar plot. Again we confirmed the results from the revtokamap that the structures of the laminar zone, as seen in Fig. 3.10, are given by the stable and unstable manifolds similar to the revtokamap. The fingers of the Poincaré plot have very large connection lengthes compared to the areas in between, caused by the manifolds, which have infinite connection lengthes. Due to the deep penetration of the manifolds into the ergodic zone, the hot plasma is connected to the wall along the manifolds and therefore along the fingers.

Finally, we extended the DED model to the real toroidal geometry and included particles and their drift effects, described relativistically. Starting with the general relativistic form of the Hamiltonian for a charged particle in an electromagnetic field, we applied a guiding center transformation to eliminate the fast gyration of the particles around the field lines. In Sec. 4.3 the Hamiltonian and the equations of motion, using the toroidal angle as the independent variable, for the guiding center drift were derived. The model for the toroidal main field in TEXTOR, including toroidal corrections like the Shafranov shift, was introduced, according to [5, 19]. Combined with the perturbation field of the DED, derived from the cylindrical

Chapter 5. Summary and conclusion

DED model, a four dimensional mapping procedure was constructed in Sec. 4.7 for the drift of relativistic particles in a time dependent DED field. Assuming that the DED operates statically, the four dimensional mapping reduces to the already well known two dimensional case. The spectrum of the perturbation and the transit frequencies of the toroidal main field can no longer ba calculated analytically. They have to be determined numerically.

The drift effects of high energetic relativistic electrons were studied. The shift of the surfaces with constant safety factor, caused by the drift, were discussed on the basis of Figs. 4.2-4.5. The surfaces of co-passing electrons are shifted outwards with increasing kinetic energy, while the surfaces of counter-passing electrons are shifted inwards. From the Hamiltonian equations of motion an ordinary differential equation was derived, which describes the developing of surfaces of constant safety factor in the (θ, r)-plane. The comparison with the mapping results showed excellent agreement.

Including the perturbation, the developing of the chaotic plasma edge with increasing kinetic particle energy was shown for co- and counter-passing electrons. Generally, particles with high kinetic energy do not really feel the chaotic magnetic field, while low energetic particles are mainly following the field lines. The very high energetic electrons with kinetic energies larger then about 10 MeV are totally confined inside the plasma. The last intact drift surface is located directly beneath the wall. This is a high energy effect, because the electrons are moving so fast inside the torus that they cannot react on the chaotic magnetic field. This can be clearly seen in the Figs. 4.17 and 4.19 for the 10 MeV and 15 MeV counter-passing electrons, respectively, which are shifted towards the coils, where the perturbation field is much larger. Nevertheless, they show a very regular behavior. This result was quantitatively confirmed by the analysis of the dependency of the escape rates on the kinetic energy. It was shown that the characteristic parameter λ for the exponential decay decreases linearly with increasing kinetic energy. The critical energy level for the confinement of the ergodic zone was extrapolated. It was also shown that λ increases with the square of the difference between the perturbation current and the critical perturbation, which was also extrapolated. We concluded that generally λ scales linearly with the corresponding energy. For the results on high energetic electrons see Refs. [37, 38]

At the typical plasma core temperatures most of the particles have kinetic energies much less than the rest energy, so that they can be described non-relativistically. In Sec. 4.11 the non-relativistic limit was derived from the general relativistic equations. The non-relativistic Hamiltonian and the equations of motion are shown, which are identical to the results of the non-relativistic model [20].

We considered a measurement of the development of the heat flux pattern at the divertor plates with changing edge safety factor, shown in Fig. 4.24, performed by M. Jakubowski at the TEXTOR-DED in Jülich. In Fig. 4.26 it is shown that the structures of the heat flux measurement can be calculated theoretically. We used the structures of laminar plots at the wall, which were calculated for field lines, according to the result that thermal particles are following the field lines. It was explained, that the structures of the laminar plots, which are related to the stable and unstable manifolds, are corresponding directly to the heat flux pattern. From the analysis of the Poincaré plots and the manifolds for several cases with different edge safety factors, we concluded that each strike zone is related to one resonance of the Poincaré section. This resonance is the last remaining island chain at the edge of the ergodic zone for the respective q_a value. This resonance dominates all other resonances below. Decreasing the edge safety factor shifts the resonances towards the wall, whereby they are destroyed. So, the next resonance becomes dominant, which results in a new strike zone. Which strike zone belongs to which resonance was determined by the analysis of the stable and unstable manifolds and is shown in Fig. 4.33. The overlapping of the strike zones was explained. When the next resonance becomes dominant, most of the particles hit the wall at the corresponding strike zone, but the effects of the resonance above, although it can no longer be observed in the Poincaré plot, have not completely vanished. These small influences vanish not until the value of q_a has reached the q-value of the respective resonance. We explained that the outer branches of the strike zones are the dominant ones. This fact is related to the different basins of attraction of the certain manifolds, which is also reflected in the numerical problem of finding all manifolds of one island chain. Only the combination of laminar plots with stable and unstable manifolds can reveal the dominant structures. The heat flux analysis is also shown and further continued in Refs. [39, 40, 41]

All necessary numerical codes were developed and implemented in C++ within the framework of this thesis. This includes the mapping codes for the creation of the Poincaré plots, the codes for the laminar plots, the codes for the determination of periodic points and their stable and unstable manifolds and the codes for the calculation of the statistical properties for all presented models. Also a code for the explicitly time depending four dimensional mapping was developed and implemented, although no results were shown here. For the visualization of the colored contour plots, like laminar plots, the MATLAB software was used.

In summary, we explained the formation of chaotic layers, the transport mechanisms within chaotic layers and between neighboring ones, using the concept of the stable and unstable manifolds. It was shown that the wall patterns and the heat and particle transport is related to the manifolds and can be explained by them. We showed and studied the drift effects of high energetic relativistic electrons in comparison to the field line dynamics. From this analysis we concluded that the low energetic particles are following the field line dynamics. Using these results, we were able to calculate, analyze and explain the structures of measurements of heat flux patterns at the divertor plates of the TEXTOR-DED fusion experiment.

Appendix

A Discrete Schrödinger map

The stable and unstable manifolds can be used for various applications. In every type of map that includes unstable objects like hyperbolic points, the dynamics are dominated by the manifolds of these unstable objects. In this thesis we only considered maps which represent the Poincaré plot of magnetic field lines, described by a Hamiltonian system. Now we consider a map, induced by the discrete nonlinear Schrödinger equation (DNS) [24]. Using the discrete Schrödinger map, called DNS map, we can construct stationary solitary solutions of the DNS by analyzing the stable and unstable manifolds of the DNS map.

The discrete Schrödinger map

$$i\partial_t \psi_j + \psi_{j+1} - 2\psi_j + \psi_{j-1} + (\sigma+1)|\psi_j|^{2\sigma}\psi_j = 0 \tag{A.1}$$

is the discrete form of the well known nonlinear Schrödinger equation [42, 43, 44]

$$i\partial_t \psi + \partial_x^2 \psi + (\sigma+1)|\psi|^{2\sigma}\psi = 0 \, . \tag{A.2}$$

We are only interested in stationary solutions of Eq. (A.1). Therefore, we use the ansatz

$$\psi_j = G_j e^{i\lambda t} \tag{A.3}$$

on Eq. (A.1) and obtain for the stationary DNS

$$G_{j+1} - 2G_j + G_{j-1} = \lambda G_j - (\sigma+1)|G_j|^{2\sigma}G_j \, . \tag{A.4}$$

Using the definitions

$$\theta_n = G_n \, , \qquad J_n = \theta_n - \theta_{n-1} \, , \tag{A.5}$$

Appendix A. Discrete Schrödinger map

we can rewrite Eq. (A.4) to some sort of generalized form of standard map

$$J_{n+1} = J_n + f(\theta_n) \quad \text{(A.6)}$$
$$\theta_{n+1} = \theta_n + J_{n+1} \quad \text{(A.7)}$$

with

$$f(\theta_n) = \lambda \theta_n - (\sigma + 1)|\theta_n|^{2\sigma} \theta_n \ . \quad \text{(A.8)}$$

Here λ is the control parameter, while σ specifies the degree of nonlinearity. In the following, we only consider $\sigma = 1$. The existence and the shape of solitary solutions for the stationary DNS (A.4) are shown in [24]. Here we will derive these solutions from the DNS map (A.6)-(A.7), using the stable and unstable manifolds.

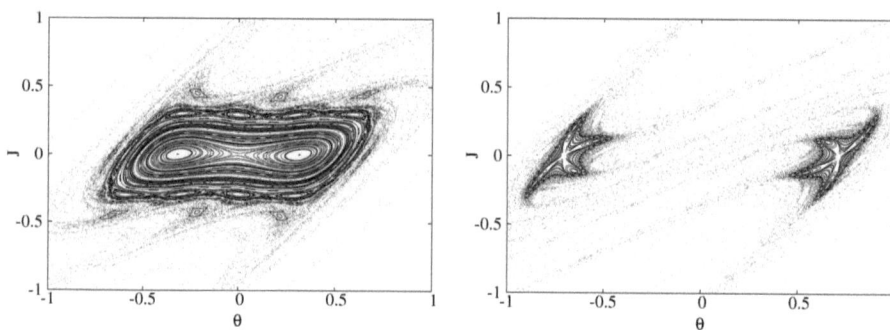

Figure 1: DNS map for $\lambda = 0.2$ Figure 2: DNS map for $\lambda = 1$

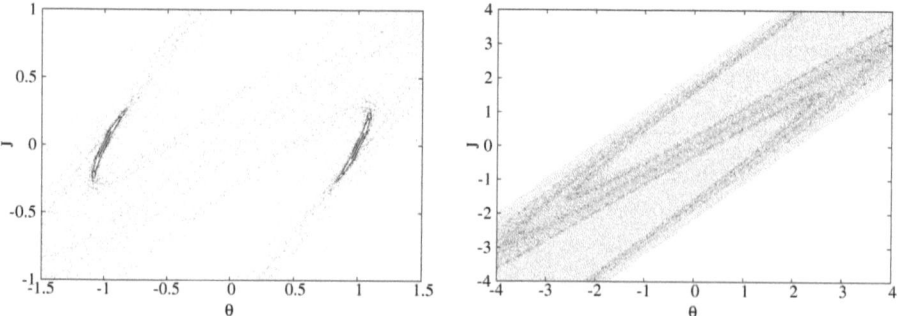

Figure 3: DNS map for $\lambda = 2$ Figure 4: DNS map for $\lambda = 4$

First, we analyze the DNS map, regarding its dependency on the control parameter λ. As one can see from Fig. 1, the DNS map has two islands on the $J = 0$ axis for small values

Appendix A. Discrete Schrödinger map

of λ. The map is symmetrical with respect to the center and shows chaotic behavior outside the islands. Between the islands directly in the center $(0,0)$ of the map, there is a period 1 hyperbolic point, which will be of great interest to us in the following. For larger values of λ, $\lambda \geq 1$, the islands vanish, as shown in Figs. 2 and 3. At $\lambda = 4$ the islands do not exist any longer, as one can see in Fig. 4, but the hyperbolic point at the center remains and characterizes the dynamics of the system. To be more precise, the stable and unstable manifolds of the hyperbolic point characterize the dynamics, because all points of the system are following the unstable manifold away from the hyperbolic point, while the stable one drives the points closer to the unstable one.

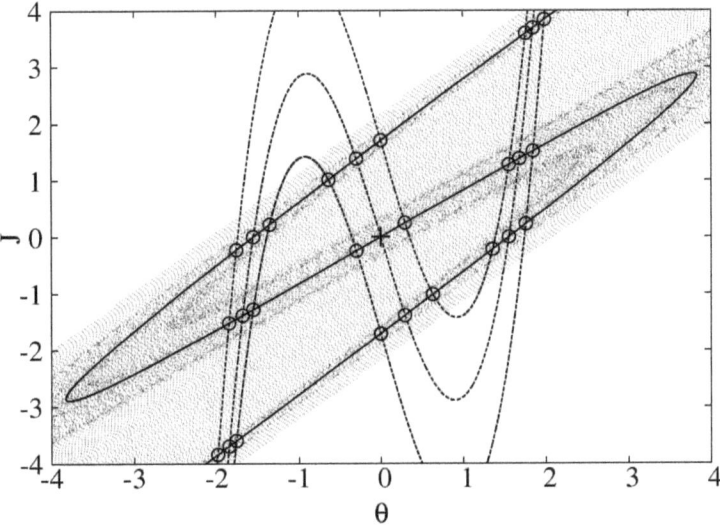

Figure 5: DNS map for $\lambda = 4$ with unstable (solid line) und stable (dashed line) manifold of the hyperbolic point at the center (cross). Intersections of the manifolds are shown by the circles.

Figure 5 shows the stable and unstable manifold of the hyperbolic point at the center and their intersections. Here only 26 intersection points are marked additionally to the fixed point, because the manifolds are plotted only up to a certain length. In fact the manifolds have infinite lengthes and there is an infinite number of intersections between the stable and unstable manifold. All iterations, backwards or forwards, of an intersection point are again intersection points, because they must be part of the stable and the unstable manifold. These intersection points are specifying the solitary solutions of the stationary DNS. Due to the definitions of the manifolds, all forward iterations converge against the hyperbolic point along the stable

Appendix A. Discrete Schrödinger map

manifold, while on the other hand all backward iterations also converge against the hyperbolic point along the unstable manifold. Thus, we can construct puls shaped solitary solutions of the stationary DNS, while all points of the solutions are intersection points of the manifolds. There can be no further solitary solution of the stationary DNS, because other points of the DNS map do not converge towards the center for $t \to \infty$ and $t \to -\infty$.

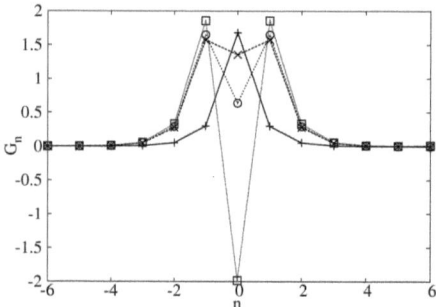 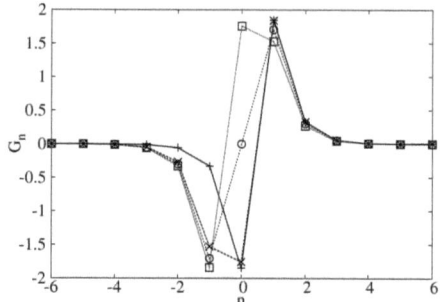

Figure 6: Four different solitary solutions of the stationary DNS with even parity at $\lambda = 4$

Figure 7: Four different solitary solutions of the stationary DNS with uneven parity at $\lambda = 4$

Figure 6 shows four different solitary solutions of the stationary DNS, constructed by the iterations of intersection points, as described above. They are all symmetrical to the y-axis at $n = 0$ and have even parity, but there are also more solutions with even parity, which are not shown here. Figure 7 shows also four different solitary solutions, but now with uneven parity. Some of the solutions presented here are also shown in [24].

Using the method described here, one can construct solitary solutions of discrete equations, as long as this equation can be rewritten to a map, e.g. (A.6)-(A.7), and this map has a hyperbolic point at the center. Here we have also shown the importance of the stable and unstable manifolds for other dynamical system then stochastic magnetic fields.

B Applying the Poisson summation rule

Here we perform the conversion of Eq. (3.1.7)

$$J = \frac{8I_0}{\theta_c r_c} g(\theta) \sum_{j=-\infty}^{\infty} \sin(\frac{j}{2}\pi + \omega t)\delta(j-n) \tag{B.1}$$

into Eq. (3.1.9)

$$J = \frac{8I_0}{\theta_c r_c} g(\theta) \sum_{k=0}^{\infty} \sin((-1)^k \frac{1}{2}\pi(2k+1)n + \omega t) , \tag{B.2}$$

as mentioned in Sec. 3.1. For this we use the Poisson summation rule (3.1.8). We begin with a straightforward manipulation of the sum over j of Eq. (B.1)

$$\sum_{j=-\infty}^{\infty} \sin(\frac{j\pi}{2} + \omega t)\delta(j-n) = \sum_j \sin(\frac{j\pi}{2})\cos(\omega t)\delta(j-n) + \sum_j \cos(\frac{j\pi}{2})\sin(\omega t)\delta(j-n)$$

$$= \sum_{k,\, j=2k+1} (-1)^k \cos(\omega t)\delta(j-n) + \sum_{k,\, j=2k} (-1)^k \sin(\omega t)\delta(j-n)$$

$$= \cos(\omega t)\sum_k \delta\big((-1)^k((2k+1)-n)\big) + \sin(\omega t)\sum_k \delta\big((-1)^k(2k-n)\big)$$

$$= \cos(\omega t)\left[\sum_{l,\, k=2l} \delta(4l+1-n) - \sum_{l,\, k=2l+1} \delta(4l+3-n)\right]$$

$$+ \sin(\omega t)\left[\sum_{l,\, k=2l} \delta(4l-n) - \sum_{l,\, k=2l+1} \delta(4l+2-n)\right]$$

$$= \frac{1}{4}\cos(\omega t)\left[\sum_l \delta(l - \frac{n-1}{4}) - \sum_l \delta(l - \frac{n-3}{4})\right]$$

$$+ \frac{1}{4}\sin(\omega t)\left[\sum_l \delta(l - \frac{n}{4}) - \sum_l \delta(l - \frac{n-2}{4})\right] =: T .$$

Now we can apply the Poisson summation rule

$$\sum_{j=-\infty}^{\infty} \delta(j-n) = 1 + 2\sum_{k=1}^{\infty} \cos(2\pi k n) \tag{B.3}$$

Appendix B. Applying the Poisson summation rule

to the preliminary result T and obtain

$$\begin{aligned} T &= \frac{1}{4}\cos(\omega t)\left[1+2\sum_{p=1}^{\infty}\cos(2\pi p\frac{n-1}{4})-1-2\sum_{p=1}^{\infty}\cos(2\pi p\frac{n-3}{4})\right] \\ &+\frac{1}{4}\sin(\omega t)\left[1+2\sum_{p=1}^{\infty}\cos(2\pi p\frac{n}{4})-1-2\sum_{p=1}^{\infty}\cos(2\pi p\frac{n-2}{4})\right] \\ &= \frac{1}{2}\cos(\omega t)\left[\sum_{p}\left(\cos(\frac{1}{2}\pi pn)\cos(\frac{1}{2}\pi p)+\sin(\frac{1}{2}\pi pn)\sin(\frac{1}{2}\pi p)\right)\right. \\ &\left. -\sum_{p}\left(\cos(\frac{1}{2}\pi pn)\cos(\frac{1}{2}\pi p)-\sin(\frac{1}{2}\pi pn)\sin(\frac{1}{2}\pi p)\right)\right] \\ &+\frac{1}{2}\sin(\omega t)\left[\sum_{p}\cos(\frac{1}{2}\pi pn)+\sum_{p}(-1)^{p+1}\cos(\frac{1}{2}\pi pn)\right]. \end{aligned}$$

Here we used $\cos(\frac{3}{2}\pi p)=\cos(\frac{1}{2}\pi p)$ and $\sin(\frac{3}{2}\pi p)=-\sin(\frac{1}{2}\pi p)$. We obtain further

$$\begin{aligned} T &= \cos(\omega t)\left[\sum_{p=1}^{\infty}\sin(\frac{1}{2}\pi pn)\sin(\frac{1}{2}\pi p)\right]+\sin(\omega t)\sum_{k=0,\,p=2k+1}^{\infty}\cos(\frac{1}{2}\pi(2k+1)n) \\ &= \cos(\omega t)\left[\sum_{k=0,\,p=2k+1}^{\infty}(-1)^{k}\sin(\frac{1}{2}\pi(2k+1)n)\right]+\sin(\omega t)\sum_{k=0,\,p=2k+1}^{\infty}\cos(\frac{1}{2}\pi(2k+1)n) \\ &= \sum_{k=0}^{\infty}\left(\cos(\omega t)\sin((-1)^{k}\frac{1}{2}\pi(2k+1)n)+\sin(\omega t)\cos((-1)^{k}\frac{1}{2}\pi(2k+1)n)\right) \\ &= \sum_{k=0}^{\infty}\sin((-1)^{k}\frac{1}{2}\pi(2k+1)n+\omega t). \end{aligned}$$

Inserting this result into Eq. (B.1), we finally find Eq. (B.2).

C Equivalent form of the DED map's generating function

There is another representation of the generating function S, which is equivalent to Eq. (3.5.6), but easier to handle with respect to derivation, and similar to the form (2.2.17), derived commonly from the mapping technique in Sec. 2.2. It reads

$$S = -\sum_m G_m(\psi)\left[h_1(z)\sin(y) + h_2(z)\cos(y)\right] \tag{C.1}$$

with the shortcuts

$$z := m\Omega - n_0, \quad y =: m\theta - n_0\varphi + \omega t, \quad k = \varphi - \varphi_0 \tag{C.2}$$

and

$$h_1(z) := \frac{\sin(kz)}{z} \to k \quad \text{for } z \to 0, \qquad h_2(z) = \frac{\cos(kz) - 1}{z} \to 0 \quad \text{for } z \to 0. \tag{C.3}$$

For the derivatives we get

$$\frac{\partial S}{\partial \theta} = \sum_m m G_m(\psi)\left[-h_1(z)\cos(y) + h_2(z)\sin(y)\right]$$

$$\frac{\partial S}{\partial \psi} = \sum_m m G_m(\psi)\left\{-\left[\frac{h_1(z)}{2\psi} + \Omega' h_1'(z)\right]\sin(y) - \left[\frac{h_2(z)}{2\psi} + \Omega' h_2'(z)\right]\cos(y)\right\}$$

$$\frac{\partial^2 S}{\partial \psi \partial \theta} = \sum_m m^2 G_m(\psi)\left\{-\left[\frac{h_1(z)}{2\psi} + \Omega' h_1'(z)\right]\cos(y) + \left[\frac{h_2(z)}{2\psi} + \Omega' h_2'(z)\right]\sin(y)\right\}$$

with

$$h_1'(z) = \frac{k\cos(kz) - h_1(z)}{z} \to 0 \quad \text{for } z \to 0 \tag{C.4}$$

$$h_2'(z) = -kh_1(z) - \frac{h_2(z)}{z} \to -\frac{1}{2}k^2 \quad \text{for } z \to 0, \tag{C.5}$$

and for the second derivatives we get

$$\frac{\partial^2 S}{\partial \psi^2} = \sum_m m G_m(\psi)\left\{\left[-\frac{m-2}{4\psi^2}h_1(z) + \left(-\frac{m}{\psi}\Omega' - \Omega''\right)h_1'(z) - m\Omega'^2 h_1''(z)\right]\sin(y)\right.$$

$$\left. + \left[-\frac{m-2}{4\psi^2}h_2(z) + \left(-\frac{m}{\psi}\Omega' - \Omega''\right)h_2'(z) - m\Omega'^2 h_2''(z)\right]\cos(y)\right\} \tag{C.6}$$

$$\frac{\partial^2 S}{\partial \theta^2} = \sum_m m^2 G_m(\psi)[h_1(z)\sin(y) + h_2(z)\cos(y)] \tag{C.7}$$

Appendix C. Equivalent form of the DED map's generating function

with

$$h_1''(z) = -k^2 h_1(z) - 2\frac{h_1'(z)}{z} \to -\frac{1}{3}k^3 \quad \text{for } z \to 0 \tag{C.8}$$

$$h_2''(z) = -k h_1'(z) + k\frac{h_1(z)}{z} + 2\frac{h_2(z)}{z^2} \to 0 \quad \text{for } z \to 0 \ . \tag{C.9}$$

Note, that y and k are defined in another way as in Sec. 3.5. This form of the generating function is used in the literature by S. Abdullaev [4, 5], but it is totally equivalent to the form used in this thesis.

Bibliography

[1] K. H. Spatschek, Einführung in die physikalischen Grundlagen der Theoretischen Astrophysik, Teubner, 2003.

[2] K. H. Finken and G. H. Wolf, Background, motivation, concept and scientific aims for building a dynamic ergodic divertor, Fusion Eng. Design 37, 337 (1997).

[3] B. Giesen, H. Bohn, W. Huettemann, O. Neubauer, M. Poier, and W. Schalt, Technical lay-out of the dynamic ergodic divertor, Fusion Eng. Design 37, 341 (1997).

[4] S. S. Abdullaev, K. H. Finken, A. Kaleck, and K. H. Spatschek, Twist Mapping for the dynamics of magnetic field lines in a tokamak ergodic divertor, Phys. Plasmas 5, 196 (1998).

[5] S. S. Abdullaev, K. H. Finken, and K. H. Spatschek, Asymptotical and mapping methods in study of ergodic divertor magnetic field in a toroidal system, Phys. Plasmas 6, 153 (1999).

[6] T. E. Evans et al., Suppression of Large Edge-Localized Modes in High-Confinement DIII-D Plasmas with a Stochastic Magnetic Boundary, Phys. Rev. Lett. 92, 235003 (2004).

[7] K. H. Finken et al., Toroidal Plasma Rotation Induced by the Dynamic Ergodic Divertor in the TEXTOR Tokamak, Phys. Rev. Lett. 94, 015003 (2005).

[8] S. S. Abdullaev, On mapping models of field lines in a stochastic magnetic field, Nuclear Fusion 44(6), 12 (2004).

[9] S. S. Abdullaev, The Hamilton-Jacobi method and Hamiltonian maps, J. Phys. A: Math. Gen. 35, 2811 (2002).

[10] S. S. Abdullaev, A new integration method of Hamiltonian systems by symplectic maps, J. Phys. A: Math. Gen. 32, 2745 (1999).

[11] R. Balescu, M. Vlad, and F. Spineanu, Tokamap: A Hamiltonian twist map for magnetic field lines in a toroidal geometry, Phys. Rev. E 58, 951 (1998).

[12] E. Nusse and J. A. Yorke, Dynamics: Numerical explorations, Springer, 1998.

[13] T. E. Evans, R. K. Roeder, J. A. Carter, and B. I. Rapoport, Homoclinic tangles, bifurcations and edge stochasticity in diverted tokamaks, Contrib. Plasma Phys. 44, 235 (2004).

[14] T. E. Evans, R. K. Roeder, J. A. Carter, B. I. Rapoport, M. E. Fenstermacher, and C. J. Lasnier, Experimental signatures of homoclinic tangles in poloidally diverted tokamaks, J. Phys.: Conf. Ser. 7, 174 (2005).

[15] R. Balescu, Hamiltonian nontwist map for magnetic field lines with locally reversed shear in toroidal geometry, Phys. Rev. E 58, 3781 (1998).

[16] J. H. Misguich, Dynamics of chaotic magnetic lines: Intermittency and noble internal transport barriers in the tokamap, Phys. Plasmas 8, 2132 (2001).

[17] J. H. Misguich, J.-D. Reuss, D. Constantinescu, G. Steinbrecher, M. Vlad, F. Spineanu, B. Weyssow, and R. Balescu, Noble Cantor sets acting as partial internal transport barriers in fusion plasmas, Plasma Phys. Contr. Fusion 44, L29 (2002).

[18] A. Wingen, K. H. Spatschek, and S. Abdullaev, Stochastic Transport of Magnetic Field Lines in the Symmetric Tokamap, Contrib. Plasma Phys. 45, 500 (2005).

[19] S. S. Abdullaev and K. H. Finken, Hamiltonian guiding center equations in a toroidal system, Phys. Plasmas 9, 4193 (2002).

[20] S. S. Abdullaev, A. Wingen, and K. H. Spatschek, Mapping of drift surfaces in toroidal systems with chaotic magnetic fields, Submitted to Phy. Plasmas (2005).

[21] H. Ali, A. Punjabi, A. Boozer, and T. Evans, The low MN map for single-null divertor tokamaks, Phys. Plasmas 11, 1908 (2004).

[22] A. Punjabi, A. Verma, and A. Boozer, Stochastic broadening of the separatrix of a tokamak divertor, Phys. Rev. Lett. 69, 3322 (1992).

[23] D. Lesnik and K. H. Spatschek, Angular transport in a nonperiodic Chirikov-Taylor map, Phys Rev E 64, 056205 (2001).

[24] E. W. Laedke, O. Kluth, and K. H. Spatschek, Existence of solitary solutions in nonlinear chains, Phys Rev E 54, 4299 (1996).

[25] R. Balescu, Matter out of Equilibrium, Imperial College Press, 1977.

[26] F. Kuypers, Klassische Mechanik, Wiley, 1997.

[27] M. Eberhard, Europhysics Conference Abstracts 23J, 781 (1999).

[28] H. H. Hasegawa and W. C. Saphir, Unitarity and irreversibility in chaotic systems, Phys. Rev. A 46, 7401 (1992).

[29] J. H. Misguich, J.-D. Reuss, D. Constantinescu, G. Steinbrecher, M. Vlad, F. Spineanu, B. Weyssow, and R. Balescu, Noble internal transport barriers and radial subdiffusion of toroidal magnetic lines, Annales Physique 28 6, 87 (2003).

[30] G. Steinbrecher, J.-D. Reuss, and J. H. Misguich, Numerical methods for finding periodic points in discrete maps: high order island chains and noble barriers in toroidal magnetic configurations, EUR-CAE-FC Report , 1719 (Nov. 2001).

[31] S. Bleher, C. Grebogi, and E. Ott, Bifurcation to chaotic scattering, Physica D 46, 87 (1990).

[32] S. S. Abdullaev, T. Eich, and K. H. Finken, Fractal structure of the magnetic field in the laminar zone of the Dynamic Ergodic Divertor of the Torus Experiment for Technology-Oriented Research (TEXTOR-94), Phys. Plasmas 8, 2739 (2001).

[33] F. Nguyen, P. Ghendrih, and A. Samain, Calculation of the magnetic field topology of ergodized edge zone in real tokamak geometry. Application to the tokamak Tore Supra through the Mastoc code, EUR-CEA-FC-1539 (1995).

[34] J. D. Jackson, Classical Electrodynamics, Wiley, 1999.

[35] M. Abramowitz and I. Stegun, Handbook of mathematical functions, Dover, 1972.

[36] W. H. Press, S. A. Teukolsky, W. T. Vetterling, and B. P. Flannery, Numerical Recipes in C, Cambridge University Press, 2002.

[37] A. Wingen, S. S. Abdullaev, K. H. Finken, and K. H. Spatschek, Influence of stochastic magnetic fields on relativistic electrons, Nucl. Fusion 46, 941 (2006).

[38] K. H. Finken, S. S. Abdullaev, M. W. Jakubowski, R. Jaspers, M. Lehnen, R. Schlickeiser, K. H. Spatschek, A. Wingen, R.Wolf, and the TEXTOR team, Runaway losses in ergodized plasmas, Nucl. Fusion 47, 91 (2007).

[39] A. Wingen, M. W. Jakubowski, K. H. Spatschek, S. S. Abdullaev, K. H. Finken, M. Lehnen, and the TEXTOR-team, Traces of stable and unstable manifolds in heat flux patterns, Phys. Plasmas 14, 042502 (2007).

[40] A. Wingen, K. H. Spatschek, S. S. Abdullaev, and M. Jakubowski, Interpretation of heat losses from open chaotic systems, Physics AUC 17, 44–58 (2007).

[41] M. W. Jakubowski, A. Wingen, S. S. Abdullaev, K. H. Finken, M. Lehnen, K. H. Spatschek, R. C. Wolf, and the TEXTOR team, Observation of the heteroclinic tangles in the heat flux pattern of the ergodic divertor at TEXTOR, J. Nucl. Mater. 363-365, 371–376 (2006).

[42] A. Hasegawa, Solitons in optical fibers, Springer, 1989.

[43] A. Hasegawa and F. D. Tappert, Transmission of stationary nonlinear optical pulses in dispersive dielectric fibers. I. Anomalous dispersion, Appl. Phys. Lett. 23, 142 (1973).

[44] A. Wingen, K. H. Spatschek, and S. B. Medvedev, Averaged dynamics of optical pulses described by a nonlinear Schrödinger equation with periodic coefficients, Phys. Rev. E 68, 046610 (2003).

Die VDM Verlagsservicegesellschaft sucht für wissenschaftliche Verlage abgeschlossene und herausragende

Dissertationen, Habilitationen, Diplomarbeiten, Master Theses, Magisterarbeiten usw.

für die kostenlose Publikation als Fachbuch.

Sie verfügen über eine Arbeit, die hohen inhaltlichen und formalen Ansprüchen genügt, und haben Interesse an einer honorarvergüteten Publikation?

Dann senden Sie bitte erste Informationen über sich und Ihre Arbeit per Email an *info@vdm-vsg.de*.

Sie erhalten kurzfristig unser Feedback!

VDM Verlagsservicegesellschaft mbH
Dudweiler Landstr. 99   Telefon +49 681 3720 174
D - 66123 Saarbrücken   Fax   +49 681 3720 1749
www.vdm-vsg.de

Die VDM Verlagsservicegesellschaft mbH vertritt

Printed by Books on Demand GmbH, Norderstedt / Germany